SUBSTITUTIONAL ANALYSIS

DANIEL EDWIN RUTHERFORD
UNIVERSITY OF ST. ANDREWS

DOVER PUBLICATIONS, INC.
MINEOLA, NEW YORK

Bibliographical Note

This Dover edition, first published in 2013, is an unabridged republication of the work originally published by Hafner Publishing Company, New York, in 1968, which was itself a republication of the edition published by Edinburgh University Press in 1948.

Library of Congress Cataloging-in-Publication Data

Rutherford, D. E. (Daniel Edwin), 1906–
Substitutional analysis / Daniel Edwin Rutherford, University of St. Andrews.
p. cm.
Originally published: Edinburgh: Edinburgh University Press, 1948.
Includes bibliographical references and index.
ISBN-13: 978-0-486-49120-2
ISBN-10: 0-486-49120-X
1. Group theory. I. Title.

QA171.R8 2013
512'.2—dc23

2012038899

Manufactured in the United States by Courier Corporation
49120X01
www.doverpublications.com

CONTENTS

PREFACE

In the preparation of this book, I have been much indebted to various colleagues and friends. In particular I have received much helpful advice and constructive criticism over a period of years from Professor H. W. Turnbull, F.R.S., and Dr. W. Ledermann, both of whom have also assisted me in proof reading. I also wish to express my grateful thanks to Professor Sir Edmund Whittaker, F.R.S., and the Edinburgh University Press for their cooperation and invaluable assistance in the production of the book.

D. E. Rutherford

August 1947

INTRODUCTION

THE purpose of this book is to give an account of the methods employed by Alfred Young in his reduction of the symmetric group and to describe the more important results achieved by him.

The problem which first attracted Young's attention and which initiated the theory developed by him was that of solving certain substitutional equations which arose in his study of the Theory of Invariants. Although this initial problem was never far from his mind, his researches led him to study problems whose significance was deeper than he originally suspected. The keystone of these was the reduction of the symmetric group to its irreducible representations and the presentation of these representations in an explicit form. His published researches on these subjects, extending from 1900 to 1935, reveal some interesting facts. Most remarkable perhaps is the gap of twenty-five years between the second paper in 1902 and the third in 1927. In the first two papers Young had introduced the concept of a tableau which is so fundamental in the subsequent theory and had achieved some interesting results. It seems fairly certain that this brilliant inspiration was arrived at by a close study of the Gordan–Capelli series in the Theory of Invariants. This is borne out by the fact that the first use he made of his newly forged tool was its application to the Gordan–Capelli series.

In the introduction to his third paper Young writes: " When writing the two former papers I suffered from the disadvantage of being unacquainted with the closely related researches of the late Prof. Frobenius, published in the Berliner *Sitzungsberichte*, and beginning with 'Ueber Gruppencharaktere', 1896; a lucid and more elementary exposition of the main features of Frobenius's theory of group characters was given by Schur ". It is uncertain at what date Young's attention was drawn to the work of Frobenius and Schur, but it is certain that their papers made a great impression on him and spurred him on to develop his own approach to the subject. To enable him to assimilate the papers of Frobenius and Schur he undertook a study of the German language. When one remembers in addition that Young was not a professional mathematician but a country clergyman with numerous clerical duties, the gap of twenty-five years between his second and third papers is not so surprising.

The third paper, when it appeared, was not so much an application of the work of Frobenius and Schur to Young's previous work as a new and valuable approach to the construction of the irreducible representations of the symmetric group. The essential feature of this new development is the selection from all possible tableaux of a sub-system of what are called standard tableaux. The results achieved in this paper enable certain types of substitutional equations to be reduced to a system of matrix equations. These matrix equations can in general be solved, although the manipulation in some cases may be enormous.

The fourth paper continues the general development of the third and provides a recipe for writing down the matrices of the natural irreducible representations of the symmetric group. Associated with any irreducible representation there is an infinite number of equivalent representations of which the natural representation is only one. Two others of particular importance, the semi-normal representation and the orthogonal representation, are characterised by the peculiar nature of the invariant quadratics associated with them. These two are obtained in Young's sixth paper, and the theory is extended still further in the eighth paper.

The fifth paper is concerned with the hyper-octahedral group. This group is treated in much the same way as the symmetric group.

Most of the papers mentioned above have large sections devoted to the applications of the theory to the theory of invariants, and certain other papers listed in the Bibliography on page 101 are exclusively devoted to these topics. These applications, however, are somewhat technical in character and will not be treated in this book. The reader who is interested in these applications should also consult the recent papers of P. G. Molenaar.

As has so frequently happened in other branches of mathematics, the development of the theory of substitutional analysis necessitated from time to time changes in the notation employed. The lack of a comprehensive and systematic survey of Young's work using a unified notation throughout explains to some extent the comparative neglect which has been accorded to Young's researches by his contemporaries. In fact, one might say that only two significant contributions have so far been made by other authors to the substitutional theory initiated by Young. The first of these was contained in an oral remark made by von Neumann to van der Waerden and is published in the latter's *Moderne Algebra*, vol. 2. This short theorem is the link which connects Young's substitutional analysis with the Ideal-theory of Abstract Algebra. It also affords a means of simplifying some of Young's more intricate proofs. The other work mentioned is a paper published by R. M. Thrall in 1941. Adopting the point of view of abstract algebra, Thrall derives Young's orthogonal representation directly and thereby eliminates a great mass of elaborate detail which Young found necessary in constructing the orthogonal representation from the natural one.

The discoveries of Frobenius and Schur are not without significance in the theory under consideration. Indeed their results embody many of those of Young. The two theories may be regarded as parallel attacks on the same problem, but in this book the emphasis will be laid on the calculus of tableaux as applied to the symmetric group, and this particular aspect of the subject is peculiar to Young's work and to that of von Neumann, Robinson and Thrall.

Likewise there have been several successful attempts, notably those of Weyl, Murnaghan and D. E. Littlewood, to relate Young's work to other branches of Modern Algebra. These, however, have already been expounded by their several authors and will not be enlarged upon here.

In this book it was considered desirable to expound the subject in terms of Young's mathematical language because in this way the theory can be studied by a reader possessing no previous knowledge of the subject apart from those portions of the Theory of Groups and the Theory of Matrices which are familiar to all mathematicians, and because in this way only can the individuality and genius of Young be properly recognised. Nevertheless, the reader who is already familiar with Young's writings will observe that the presentation here given varies considerably from that of Young. In some cases the order of development has been changed, and in consequence of this, many of the proofs given are new. It is hoped that these changes will contribute to the lucidity and beauty of the underlying theory. It will therefore be understood that the references in the text to Young's work do not necessarily imply that the proof given is due to Young. While this is so in some cases, in others the reference is quoted only to show that the result in question was also obtained in whole or in part by Young. The references quoted in the text are given in an abridged form. The works cited are given in full in the Bibliography on page 101.

CHAPTER I

THE CALCULUS OF PERMUTATIONS

Argument. The $n!$ possible permutations of n letters occupy a place of basic importance throughout this book. In this first chapter we shall describe some of their more interesting properties and shall introduce certain notations which will facilitate our investigations in later chapters. Most readers will find that much of this chapter is already familiar to them but they should nevertheless pay particular attention to the notations employed, especially in § 6 and § 7.

§ 1. Permutations

It is well known that there are $n!$ different permutations which can be made on n letters. We shall call these letters $z_1, \ldots, z_n$, but in most cases it will be more convenient to denote them by their suffixes only. To avoid ambiguity we shall use a special fount of type for this purpose and write $z_1, \ldots, z_n$ simply as $\mathbf{1}, \ldots, \mathbf{n}$. Thus

$$\begin{pmatrix} \mathbf{1}, \mathbf{2}, \ldots, \mathbf{n} \\ \mathbf{i}_1, \mathbf{i}_2, \ldots, \mathbf{i}_n \end{pmatrix}, \qquad \ldots(1.1)$$

where $\mathbf{i}_1, \ldots, \mathbf{i}_n$ denote the letters $\mathbf{1}, \ldots, \mathbf{n}$ in some order, denotes the permutation which changes the letter $\mathbf{1}$ into the letter $\mathbf{i}_1$ and so on. There is no particular reason why the columns of this permutation should be written in any special order. The same permutation might be denoted by

$$\begin{pmatrix} \mathbf{2}, \mathbf{1}, \ldots, \mathbf{n} \\ \mathbf{i}_2, \mathbf{i}_1, \ldots, \mathbf{i}_n \end{pmatrix}$$

or, more generally, by

$$\begin{pmatrix} \mathbf{k}_1, \mathbf{k}_2, \ldots, \mathbf{k}_n \\ \mathbf{i}_{k_1}, \mathbf{i}_{k_2}, \ldots, \mathbf{i}_{k_n} \end{pmatrix},$$

where $\mathbf{k}_1, \ldots, \mathbf{k}_n$ are the letters $\mathbf{1}, \ldots, \mathbf{n}$ in some order. Greek letters $\sigma, \tau, \ldots$ will frequently be employed to denote permutations in cases where the above more precise notation is unnecessary. In particular we shall write

$$\epsilon \equiv \begin{pmatrix} \mathbf{1}, \mathbf{2}, \ldots, \mathbf{n} \\ \mathbf{1}, \mathbf{2}, \ldots, \mathbf{n} \end{pmatrix}$$

throughout this book, where ϵ denotes the identical permutation.

In Group Theory these permutations are considered as elements which have an independent existence, but for many of our purposes they must be thought of as operators which are applicable to functions of the n letters. Thus, when the permutation (1.1) is applied to a function $F(z_1, \ldots, z_n)$, a function $F(z_{i_1}, \ldots, z_{i_n})$ is obtained which is in general different from $F(z_1, \ldots, z_n)$. This is expressed by the formula

$$\begin{pmatrix} 1, 2, \ldots, n \\ i_1, i_2, \ldots, i_n \end{pmatrix} F(z_1, \ldots, z_n) = F(z_{i_1}, \ldots, z_{i_n}). \qquad \ldots(1.2)$$

For this reason we shall write the result of operating on F first with the permutation σ_1 and then with the permutation σ_2 as $\sigma_2\sigma_1F$ and not as $\sigma_1\sigma_2F$. It is easily proved that the effect of operating successively with two permutations is equivalent to operating with a single permutation, which we may call the product of the other two. In other words, $\sigma_2\sigma_1$ is itself a permutation σ_3, namely, that permutation which results from first performing the permutation σ_1, and then performing σ_2. It should be borne in mind that in Group Theory, where the permutations are not thought of as operators acting on functions, the product which we now denote by $\sigma_2\sigma_1$ is more usually written $\sigma_1\sigma_2$. This distinction is an important one since in general permutations are non-commutative with respect to multiplication. The truth of this is evident from the following illustration :

$$\begin{pmatrix} 1, 2, 3 \\ 2, 1, 3 \end{pmatrix}\begin{pmatrix} 1, 2, 3 \\ 1, 3, 2 \end{pmatrix} = \begin{pmatrix} 1, 2, 3 \\ 2, 3, 1 \end{pmatrix},$$
$$\begin{pmatrix} 1, 2, 3 \\ 1, 3, 2 \end{pmatrix}\begin{pmatrix} 1, 2, 3 \\ 2, 1, 3 \end{pmatrix} = \begin{pmatrix} 1, 2, 3 \\ 3, 1, 2 \end{pmatrix}.$$

We shall see in a moment that every permutation σ possesses an inverse σ^{-1}. Since two permutations σ_1 and σ_2 do not necessarily commute, we must be careful to write

$$(\sigma_1\sigma_2)^{-1} = \sigma_2^{-1}\,\sigma_1^{-1},$$

as is usual in all non-commutative algebra.

§ 2. The Symmetric Group $\mathcal{S}_n$

From time to time it will be necessary to quote certain standard results from the more elementary parts of the theory of finite groups. Such results are very well known and appear in all the usual text-books. This being so, it would seem to be superfluous to include the proofs of these results in this book. An exception will nevertheless be made in the case of theorems and formulae which are specifically concerned with the $n!$ permutations of the symmetric group. In such cases there will be no objection to expressing the proofs concerned in a concise form.

THEOREM 1. *The $n!$ permutations of n letters form a group.*

PROOF. (i) The product of any two permutations is a permutation.

(ii) The identity permutation

$$\epsilon \equiv \begin{pmatrix} 1, 2, \ldots, n \\ 1, 2 \; \ldots \; n \end{pmatrix}$$

which leaves every letter unaltered is the unit element of the group.

(iii) Since

$$\begin{pmatrix} \mathbf{i}_1, \ldots, \mathbf{i}_n \\ \mathbf{1}, \ldots, \mathbf{n} \end{pmatrix} \begin{pmatrix} \mathbf{1}, \ldots, \mathbf{n} \\ \mathbf{i}_1, \ldots, \mathbf{i}_n \end{pmatrix} = \begin{pmatrix} \mathbf{1}, \ldots, \mathbf{n} \\ \mathbf{1}, \ldots, \mathbf{n} \end{pmatrix} = \epsilon,$$

every permutation

$$\begin{pmatrix} \mathbf{1}, \ldots, \mathbf{n} \\ \mathbf{i}_1, \ldots, \mathbf{i}_n \end{pmatrix}$$

has an inverse

$$\begin{pmatrix} \mathbf{i}_1, \ldots, \mathbf{i}_n \\ \mathbf{1}, \ldots, \mathbf{n} \end{pmatrix}.$$

(iv) It can readily be verified that the associative law holds.

The $n!$ permutations of n letters therefore satisfy the four group postulates and consequently form a group. This group which is called the *symmetric group* of order $n!$ is usually denoted by $\mathfrak{S}_n$.

THEOREM 2. *If σ and τ be any two permutations the product $\sigma\tau\sigma^{-1}$ is that permutation which is obtained by operating* on τ with σ.*

PROOF. Let

$$\tau = \begin{pmatrix} \mathbf{1}, \ldots, \mathbf{n} \\ \mathbf{j}_1, \ldots, \mathbf{j}_n \end{pmatrix}, \quad \sigma = \begin{pmatrix} \mathbf{1}, \ldots, \mathbf{n} \\ \mathbf{i}_1, \ldots, \mathbf{i}_n \end{pmatrix} = \begin{pmatrix} \mathbf{j}_1, \ldots, \mathbf{j}_n \\ \mathbf{k}_1, \ldots, \mathbf{k}_n \end{pmatrix}.$$

Then

$$\sigma^{-1} = \begin{pmatrix} \mathbf{i}_1, \ldots, \mathbf{i}_n \\ \mathbf{1}, \ldots, \mathbf{n} \end{pmatrix}$$

and

$$\sigma\tau\sigma^{-1} = \begin{pmatrix} \mathbf{j}_1, \ldots, \mathbf{j}_n \\ \mathbf{k}_1, \ldots, \mathbf{k}_n \end{pmatrix} \begin{pmatrix} \mathbf{1}, \ldots, \mathbf{n} \\ \mathbf{j}_1, \ldots, \mathbf{j}_n \end{pmatrix} \begin{pmatrix} \mathbf{i}_1, \ldots, \mathbf{i}_n \\ \mathbf{1}, \ldots, \mathbf{n} \end{pmatrix} = \begin{pmatrix} \mathbf{i}_1, \ldots, \mathbf{i}_n \\ \mathbf{k}_1, \ldots, \mathbf{k}_n \end{pmatrix}.$$

Clearly this last permutation can be obtained from τ by operating on it with σ.

The significance of this theorem, which is of fundamental importance in our subsequent work, can perhaps best be understood by comparing the relations

$$\begin{pmatrix} \mathbf{1}, \mathbf{2}, \mathbf{3} \\ \mathbf{2}, \mathbf{1}, \mathbf{3} \end{pmatrix} \begin{pmatrix} \mathbf{1}, \mathbf{2}, \mathbf{3} \\ \mathbf{2}, \mathbf{3}, \mathbf{1} \end{pmatrix} \begin{pmatrix} \mathbf{2}, \mathbf{1}, \mathbf{3} \\ \mathbf{1}, \mathbf{2}, \mathbf{3} \end{pmatrix} = \begin{pmatrix} \mathbf{2}, \mathbf{1}, \mathbf{3} \\ \mathbf{1}, \mathbf{3}, \mathbf{2} \end{pmatrix},$$

$$\begin{pmatrix} \mathbf{1}, \mathbf{2}, \mathbf{3} \\ \mathbf{2}, \mathbf{1}, \mathbf{3} \end{pmatrix} F(z_1, z_2, z_3) = F(z_2, z_1, z_3).$$

§ 3. Cycles and Transpositions

The most fruitful way of investigating permutations is by expressing them as products of cycles. By a *cycle* is meant a permutation of the type

$$\begin{pmatrix} \mathbf{i}_1, \mathbf{i}_2, \ldots, \mathbf{i}_{r-1}, \mathbf{i}_r, \mathbf{i}_{r+1}, \ldots, \mathbf{i}_n \\ \mathbf{i}_2, \mathbf{i}_3, \ldots, \mathbf{i}_r, \quad \mathbf{i}_1, \mathbf{i}_{r+1}, \ldots, \mathbf{i}_n \end{pmatrix},$$

which leaves the letters $\mathbf{i}_{r+1}, \ldots, \mathbf{i}_n$ unaltered but which permutes the letters

* Operating in the sense of (1.2).

$i_1, \ldots, i_r$ cyclically. The number r of letters permuted cyclically is called the *order* of the cycle. Such a cycle may be economically written

$$(i_1, \ldots, i_r).$$

It might also be written $(i_2, \ldots, i_r, i_1)$, or again $(i_r, i_1, \ldots, i_{r-1})$, or in fact in r different ways in all.

The notation just introduced has the advantage that only those letters which are affected by the cycle are displayed. It is therefore clear at a glance which letters are unaffected. Thus the cycle (2, 1, 4) leaves the letter 3 unaltered. A set of cycles no two of which affect the same letter are said to be *independent*. It is an elementary fact of considerable importance that independent cycles commute with one another. This must be so since in any product of independent cycles no letter is affected more than once.

It follows from our definition that the rth power of a cycle of order r is always the unit element; thus if $\sigma = (i_1, \ldots, i_r)$, then $\sigma^r = \epsilon$ and $\sigma^{r-1} = \sigma^{-1}$. The inverse of a cycle of order r is a cycle of order r, namely that which permutes the letters $i_1, \ldots, i_r$ cyclically in the reverse order. These properties will be clarified by the following illustration:

$$\sigma = (1, 2, \ldots, r) = \begin{pmatrix} 1, 2, \ldots, r, r+1, \ldots, n \\ 2, 3, \ldots, 1, r+1, \ldots, n \end{pmatrix},$$

$$\sigma^2 = (1, 2, \ldots, r)^2 = \begin{pmatrix} 1, 2, \ldots, r, r+1, \ldots, n \\ 3, 4, \ldots, 2, r+1, \ldots, n \end{pmatrix},$$

$$\cdots\cdots\cdots\cdots\cdots\cdots\cdots\cdots\cdots\cdots\cdots\cdots$$

$$\sigma^{r-1} = (1, 2, \ldots, r)^{r-1} = \begin{pmatrix} 1, 2, \ldots, r, & r+1, \ldots, n \\ r, 1, \ldots, r-1, & r+1, \ldots, n \end{pmatrix}$$

$$= (r, r-1 \ldots, 1),$$

$$\sigma^r = (1, 2, \ldots, r)^r = \begin{pmatrix} 1, 2, \ldots, r, r+1, \ldots, n \\ 1, 2, \ldots, r, r+1, \ldots, n \end{pmatrix}$$

$$= \epsilon.$$

Every permutation can be expressed as a product of independent cycles. The method of doing so is typified by the following example:

$$\begin{pmatrix} 1, 2, 3, 4, 5, 6, 7 \\ 3, 6, 2, 7, 5, 1, 4 \end{pmatrix} = \begin{pmatrix} 1, 3, 2, 6, 4, 7, 5 \\ 3, 2, 6, 1, 7, 4, 5 \end{pmatrix}$$

$$= \begin{pmatrix} 1, 3, 2, 6, 4, 7, 5 \\ 3, 2, 6, 1, 4, 7, 5 \end{pmatrix} \begin{pmatrix} 4, 7, 1, 3, 2, 6, 5 \\ 7, 4, 1, 3, 2, 6, 5 \end{pmatrix} \begin{pmatrix} 5, 1, 3, 2, 6, 4, 7 \\ 5, 1, 3, 2, 6, 4, 7 \end{pmatrix}$$

$$= (1, 3, 2, 6)\,(4, 7)\,(5).$$

Since any cycle of order unity is merely the identity permutation ϵ we may omit such cycles from any product. The permutation illustrated above is most concisely written (1, 3, 2, 6) (4, 7).

Using this notation the 3! permutations of $\mathcal{S}_3$ are

$$\begin{pmatrix}1, 2, 3\\1, 2, 3\end{pmatrix}=\epsilon, \qquad \begin{pmatrix}1, 2, 3\\2, 3, 1\end{pmatrix}=(1, 2, 3), \quad \begin{pmatrix}1, 2, 3\\3, 1, 2\end{pmatrix}=(3, 2, 1),$$

$$\begin{pmatrix}1, 2, 3\\1, 3, 2\end{pmatrix}=(2, 3), \quad \begin{pmatrix}1, 2, 3\\3, 2, 1\end{pmatrix}=(3, 1) \qquad \begin{pmatrix}1, 2, 3\\2, 1, 3\end{pmatrix}=(1, 2).$$

The group table which tabulates all products $\sigma\tau$ of this group is given below. The elements σ are in the column on the left and the elements τ lie in the row at the top.

$\mathcal{S}_3$	ϵ	(1, 2, 3)	(3, 2, 1)	(2, 3)	(3, 1)	(1, 2)
ϵ	ϵ	(1, 2, 3)	(3, 2, 1)	(2, 3)	(3, 1)	(1, 2)
$(1, 2, 3)^{-1}=(3, 2, 1)$	(3, 2, 1)	ϵ	(1, 2, 3)	(3, 1)	(1, 2)	(2, 3)
$(3, 2, 1)^{-1}=(1, 2, 3)$	(1, 2, 3)	(3, 2, 1)	ϵ	(1, 2)	(2, 3)	(3, 1)
$(2, 3)^{-1}=(2, 3)$	(2, 3)	(3, 1)	(1, 2)	ϵ	(1, 2, 3)	(3, 2, 1)
$(3, 1)^{-1}=(3, 1)$	(3, 1)	(1, 2)	(2, 3)	(3, 2, 1)	ϵ	(1, 2, 3)
$(1, 2)^{-1}=(1, 2)$	(1, 2)	(2, 3)	(3, 1)	(1, 2, 3)	(3, 2, 1)	ϵ

A cycle of order two is called a *transposition* and is evidently its own inverse. It can be readily verified that

$$\begin{aligned}(1, 2, \ldots, r)&=(1, r)(1, r-1)\ldots(1, 3)(1, 2)\\&=(1, 2)(2, 3)\ldots(r-2, r-1)(r-1, r).\end{aligned}$$

Many other such resolutions are possible ; e.g.

$$(1, 2, \ldots, r)=(2, 3, \ldots, r, 1)=(2, 1)(2, r)\ldots(2, 3);$$

but it must be remembered that transpositions like cycles do not commute if they have letters in common. It follows that any permutation, being a product of cycles, can always be expressed as a product of transpositions. Thus,

$$\begin{pmatrix}1, 2, 3, 4, 5, 6, 7\\3, 6, 2, 7, 5, 1, 4\end{pmatrix}=(1, 3, 2, 6)(4, 7)=(1, 3)(3, 2)(2, 6)(4, 7).$$

We are now in a position to prove the following result.

THEOREM 3. *Every permutation can be expressed as a product of transpositions of the form* $(k-1, k)$ *where* $k-1$ *and* k *are two consecutive letters.*

PROOF. Since we have already expressed any permutation as a product of transpositions, it remains to show that any transposition (i, j) can be expressed as a product of transpositions of the form $(k-1, k)$. From theorem 2 (§ 2) it follows that

$$(i, j)=(1, i)(1, j)(1, i)$$

and that

$$(1, i) = (i, \ldots, 3, 2)\,(1, 2)\,(2, 3, \ldots, i).$$

Also, we have seen that

$$(i, \ldots, 3, 2) = (i, i-1) \ldots (4, 3)\,(3, 2),$$
$$(2, 3, \ldots, i) = (2, 3)\,(3, 4) \ldots (i-1, i).$$

A combination of these formulae yields the desired result.

For example,

$$(2, 4) = (1, 2)\,(1, 4)\,(1, 2) = (1, 2)\,(3, 4)\,(2, 3)\,(1, 2)\,(2, 3)\,(3, 4)\,(1, 2).$$

§ 4. Odd and Even Permutations

When any permutation σ is applied to the alternant

$$\Delta \equiv \prod_{j>i} (z_j - z_i)$$

the value of this expression remains unaltered apart from sign. Now for any given σ we must have either $\sigma\Delta = +\Delta$ or else $\sigma\Delta = -\Delta$. If $\sigma\Delta = +\Delta$ we call σ an *even permutation*, whereas if $\sigma\Delta = -\Delta$ we call σ an *odd permutation*. Clearly every transposition is an odd permutation, for when it is applied to Δ it changes the sign of an odd number of factors. It follows from this that the product of an even number of transpositions is an even permutation and that the product of an odd number of transpositions is an odd permutation. Although in general any permutation can be expressed in a variety of ways as the product of transpositions, it is clear from the foregoing that each such way involves an even number of transpositions if the permutation be an even one, but an odd number of transpositions if the permutation be odd. In illustration we remark that ϵ, $(1, 2, 3)$, $(1, 2)\,(3, 4)$ are even permutations and that $(1, 2)$, $(1, 2, 3, 4)$ are odd permutations.

Associated with each permutation σ we now define a number ζ_σ with the properties

$$\zeta_\sigma = +1 \textit{ if } \sigma \textit{ is an even permutation},$$
$$\zeta_\sigma = -1 \textit{ if } \sigma \textit{ is an odd permutation}.$$

This notation will prove very useful in subsequent chapters. It is already familiar in the theory of determinants; e.g.

$$\begin{vmatrix} x_1 \ldots x_n \\ \ldots\ldots \\ t_1 \ldots t_n \end{vmatrix} = \sum_\sigma \zeta_\sigma \sigma(x_1 \ldots, t_n).$$

§ 5. Classes of Permutations

The concept of a class of elements is a very important one in the Theory of Groups and we shall make some use of it at a later stage in this book. For our purposes it will suffice to consider the case of the symmetric group $\mathcal{S}_n$ only. Two permutations τ_1 and τ_2 of $\mathcal{S}_n$ are said to belong to the same *class* if it is possible to find a permutation σ of $\mathcal{S}_n$ such that

$$\sigma\tau_1\sigma^{-1} = \tau_2.$$

Theorem 2 (§ 2) tells us that this is possible if and only if τ_2 can be obtained by operating on τ_1 with some permutation σ. This means that τ_1 and τ_2 must be built up of the same number of independent cycles and that the orders of these component cycles are the same in each case although the arrangement of the letters in the cycles will be different when τ_1 and τ_2 are distinct. In other words, all those permutations which are the products of independent cycles of orders $\alpha_1, \ldots, \alpha_k$ form a class of $\mathcal{S}_n$ and no other permutations of $\mathcal{S}_n$ belong to this class. In particular, $\mathcal{S}_3$ has three classes, namely,

$$\epsilon; \quad (1, 2, 3), (3, 2, 1); \quad (2, 3), (3, 1), (1, 2).$$

Since we have shown that the inverse of a cycle of order r is a cycle of order r, it is patent that σ and σ^{-1} must always belong to the same class of $\mathcal{S}_n$. A corresponding statement is not true for every finite group.

If τ_1 is expressible as the product of r transpositions, then $\sigma\tau_1\sigma^{-1}$ is also expressible as the product of r transpositions. It follows from this that all the permutations of a given class are either all odd permutations or else are all even permutations.

§ 6. Substitutional Expressions

The substitutional expressions which we now introduce can be viewed from two angles. Although the permutations σ, τ considered as elements of the group $\mathcal{S}_n$ admit of only one law of combination, namely multiplication, yielding products such as $\sigma\tau$ and $\tau\sigma$, we can attach a meaning to $\sigma+\tau$ if we regard the permutations as operators acting on a function F of the n letters $z_1, \ldots, z_n$. If $\sigma F \equiv F_\sigma$, $\tau F \equiv F_\tau$, we define $\sigma+\tau$ to be that operation which yields the function $F_\sigma + F_\tau$ when it is applied to the function F. In view of this definition we can extend the above notation and write

$$F_\sigma + F_\tau \equiv F_{(\sigma+\tau)}.$$

We can generalise the foregoing idea by attaching numerical coefficients to the permutations. The general *substitutional expression* has the form

$$X \equiv \lambda_1\epsilon + \lambda_2\sigma_2 + \ldots + \lambda_{n!}\sigma_{n!},$$

where $\epsilon, \sigma_2, \ldots, \sigma_{n!}$ are the $n!$ distinct permutations of $\mathcal{S}_n$ and where $\lambda_1, \ldots, \lambda_{n!}$ are numerical coefficients. X is defined as that operation which when applied to any function F yields F_X, where

$$F_X \equiv \lambda_1 F + \lambda_2 F_{\sigma_2} + \ldots + \lambda_{n!} F_{\sigma_{n!}}.$$

and where $F_{\sigma_i} \equiv \sigma_i F$. These substitutional expressions form the main topic of this book.

To take a more abstract point of view we may define the substitutional expressions X as hyper-complex numbers constructed from the permutations σ_i of $\mathcal{S}_n$ as units. The resulting algebra is sometimes called the *group algebra* of $\mathcal{S}_n$. Any two substitutional expressions $X \equiv \sum_i \lambda_{\sigma_i}\sigma_i$ and $Y \equiv \sum_i \mu_{\sigma_i}\sigma_i$ have a sum

$$X + Y \equiv \sum_i (\lambda_{\sigma_i} + \mu_{\sigma_i})\sigma_i$$

and a product

$$XY \equiv \sum_{i,j} \lambda_{\sigma_i}\mu_{\sigma_j}\sigma_i\sigma_j.$$

If we write $\sigma_i\sigma_j = \sigma_k$, then $\sigma_j = \sigma_i^{-1}\sigma_k$; thus we may also write

$$XY \equiv \sum_{i,k} \lambda_{\sigma_i}\mu_{\sigma_i^{-1}\sigma_k}\sigma_k,$$

so that not only the sum $X + Y$ but also the product XY is a linear combination of the permutations σ_i. As such it is clear that the sum and the product of two substitutional expressions are also substitutional expressions. As illustrations of this the reader may verify, with the help of the group table for $\mathcal{S}_3$ (§ 3), that if

$$X = \epsilon - 2(\mathbf{1}, \mathbf{2}), \quad Y = 3(\mathbf{1}, \mathbf{2}, \mathbf{3}) + (\mathbf{1}, \mathbf{2}),$$

then

$$\begin{aligned} X + Y &= \epsilon - (\mathbf{1}, \mathbf{2}) + 3(\mathbf{1}, \mathbf{2}, \mathbf{3}), \\ XY &= -2\epsilon + (\mathbf{1}, \mathbf{2}) - 6(\mathbf{2}, \mathbf{3}) + 3(\mathbf{1}, \mathbf{2}, \mathbf{3}), \\ YX &= -2\epsilon + (\mathbf{1}, \mathbf{2}) - 6(\mathbf{3}, \mathbf{1}) + 3(\mathbf{1}, \mathbf{2}, \mathbf{3}). \end{aligned}$$

§ 7. The Positive and Negative Symmetric Groups on r Letters

Certain types of substitutional expressions are of frequent occurrence and as such deserve a special notation. The set of all permutations σ obtainable by permuting the letters $\mathbf{i}_1, \ldots, \mathbf{i}_r$ among themselves constitute a symmetric group of order $r!$. To be more precise, we shall call it the *positive symmetric group* on the r letters $\mathbf{i}_1, \ldots, \mathbf{i}_r$. The sum of all the elements of this positive symmetric group will be denoted by

$$\{\mathbf{i}_1, \ldots, \mathbf{i}_r\}$$

and we can denote the positive symmetric group itself by

$$\mathcal{G}\{\mathbf{i}_1, \ldots, \mathbf{i}_r\}.$$

It is easily proved that the substitutional expressions $\zeta_\sigma \sigma$, where σ ranges over the permutations of $\mathcal{G}\{\mathbf{i}_1, \ldots, \mathbf{i}_r\}$, also constitute a group which is simply isomorphic with $\mathcal{G}\{\mathbf{i}_1, \ldots, \mathbf{i}_r\}$. This group is called the *negative symmetric group* on the r letters $\mathbf{i}_1, \ldots, \mathbf{i}_r$ and will be denoted by

$$\mathcal{G}\{\mathbf{i}_1, \ldots, \mathbf{i}_r\}'.$$

The sum of its elements can be written

$$\{\mathbf{i}_1, \ldots, \mathbf{i}_r\}'.$$

From the abstract point of view $\mathcal{G}\{\mathbf{i}_1, \ldots, \mathbf{i}_r\}$ and $\mathcal{G}\{\mathbf{i}_1, \ldots, \mathbf{i}_r\}'$ are merely different representations of $\mathcal{S}_r$. The elements of these representations are substitutional expressions and in the general case $\{\mathbf{i}_1, \ldots, \mathbf{i}_r\}$ and $\{\mathbf{i}_1, \ldots, \mathbf{i}_r\}'$ are distinct elements of the group algebra. For example

$$\{1, 2, 3\} = \epsilon + (1, 2, 3) + (3, 2, 1) + (2, 3) + (3, 1) + (1, 2),$$
$$\{1, 2, 3\}' = \epsilon + (1, 2, 3) + (3, 2, 1) - (2, 3) - (3, 1) - (1, 2).$$

It is a fundamental fact in the Theory of Groups that when each element of the group is pre- or post-multiplied by some specified element of the group, the resulting products comprise all the elements of the group, each occurring once only. It follows from this that if π be any element of $\mathcal{G}\{\mathbf{i}_1, \ldots, \mathbf{i}_r\}$ and $\nu(=\zeta_\pi\pi)$ be any element of $\mathcal{G}\{\mathbf{i}_1, \ldots, \mathbf{i}_r\}'$, then

$$\pi\{\mathbf{i}_1, \ldots, \mathbf{i}_r\} = \{\mathbf{i}_1, \ldots, \mathbf{i}_r\}\pi = \{\mathbf{i}_1, \ldots, \mathbf{i}_r\},$$
$$\nu\{\mathbf{i}_1, \ldots, \mathbf{i}_r\}' = \{\mathbf{i}_1, \ldots, \mathbf{i}_r\}'\nu = \{\mathbf{i}_1, \ldots, \mathbf{i}_r\}'.$$

Here it is to be observed that if the letters $\mathbf{i}_1, \ldots, \mathbf{i}_r$ all belong to the set $\mathbf{1}, \ldots, \mathbf{n}$, then π is necessarily an element of $\mathcal{S}_n$. So is ν if it be an even element, but if it is odd then it is $-\nu$ that is an element of $\mathcal{S}_n$.

If $\mathcal{H}$ is a sub-group of order m of the group $\mathcal{G}\{\mathbf{i}_1, \ldots, \mathbf{i}_r\}$, then it is known from the Theory of Groups that elements $\pi_2, \ldots, \pi_{r!/m}$ of $\mathcal{G}\{\mathbf{i}_1, \ldots, \mathbf{i}_r\}$ can be found such that the elements of

$$\mathcal{H}, \pi_2\mathcal{H}, \ldots, \pi_{r!/m}\mathcal{H}$$

include all those of $\mathcal{G}\{\mathbf{i}_1, \ldots, \mathbf{i}_r\}$ once each and no others. Now if $\mathbf{i}_h$, $\mathbf{i}_k$ be two letters of the set $\mathbf{i}_1, \ldots, \mathbf{i}_r$, then $\mathcal{G}\{\mathbf{i}_h, \mathbf{i}_k\}$ is a sub-group of $\mathcal{G}\{\mathbf{i}_1, \ldots, \mathbf{i}_r\}$. It follows that elements $\pi_2, \ldots, \pi_{r!/2}$ can be found such that

$$\{\mathbf{i}_1, \ldots, \mathbf{i}_r\} = (\epsilon + \pi_2 + \ldots + \pi_{r!/2})\{\mathbf{i}_h, \mathbf{i}_k\}.$$

This factor $\{\mathbf{i}_h, \mathbf{i}_k\}$ is equal to $\epsilon + (\mathbf{i}_h, \mathbf{i}_k)$. It may be written either on the left or the right of the other factor $(\epsilon + \pi_2 + \ldots + \pi_{r!/2})$ as desired. A very similar argument would show that $\{\mathbf{i}_1, \ldots, \mathbf{i}_r\}'$ has a factor $\{\mathbf{i}_h, \mathbf{i}_k\}'$ or $\epsilon - (\mathbf{i}_h, \mathbf{i}_k)$. This factor can also be written either on the left or the right.

As a consequence of these facts we are able to prove an important result.

THEOREM 4. *If the sequences* $\mathbf{i}_1, \ldots, \mathbf{i}_r$ *and* $\mathbf{j}_1, \ldots, \mathbf{j}_s$ *have two letters in common, then*

$$\{\mathbf{i}_1, \ldots, \mathbf{i}_r\}\{\mathbf{j}_1, \ldots \mathbf{j}_s\}' = 0,$$
$$\{\mathbf{i}_1, \ldots, \mathbf{i}_r\}'\{\mathbf{j}_1, \ldots, \mathbf{j}_s\} = 0.$$

PROOF. Let $\mathbf{h}$, $\mathbf{k}$ be the two letters in common. Then

$$\{\mathbf{i}_1, \ldots, \mathbf{i}_r\} = [\ \ldots\ldots\ ][\epsilon + (\mathbf{h}, \mathbf{k})],$$
$$\{\mathbf{j}_1, \ldots, \mathbf{j}_s\}' = [\epsilon - (\mathbf{h}, \mathbf{k})][\ \ldots\ldots\ ].$$

The product $\{\mathbf{i}_1, \ldots, \mathbf{i}_r\}\{\mathbf{j}_1, \ldots, \mathbf{j}_s\}'$ therefore contains a factor $[\epsilon - (\mathbf{h}, \mathbf{k})^2]$ in the middle which vanishes since $(\mathbf{h}, \mathbf{k})^2 = \epsilon$. The second result may be proved in the same way.

Another result which will be required at a later stage depends upon the fact that $\mathcal{G}\{\mathbf{i}_1, \ldots, \mathbf{i}_r\}$ has a sub-group $\mathcal{G}\{\mathbf{i}_1, \ldots, \mathbf{i}_{r-1}\}$ of order $(r-1)!$. It follows that we can write

$$\{\mathbf{i}_1, \ldots, \mathbf{i}_r\} = \{\mathbf{i}_1 \ldots, \mathbf{i}_{r-1}\}(\epsilon + \pi_2 + \ldots + \pi_r),$$

where $\pi_2, \ldots, \pi_r$ are suitably chosen permutations. We can in fact choose $\pi_2, \ldots, \pi_r$ to be the transpositions $(\mathbf{i}_r, \mathbf{i}_1), \ldots, (\mathbf{i}_r, \mathbf{i}_{r-1})$, as will now be shown. Since there are exactly $r!$ terms in the expression

$$\{\mathbf{i}_1, \ldots, \mathbf{i}_{r-1}\}(\epsilon + (\mathbf{i}_r, \mathbf{i}_1) + \ldots + (\mathbf{i}_r, \mathbf{i}_{r-1}))$$

and since each of them is a term of $\{\mathbf{i}_1, \ldots, \mathbf{i}_r\}$, it has only to be shown that these $r!$ terms are all different. Suppose this were not the case and that

$$\sigma(\mathbf{i}_r, \mathbf{i}_j) = \tau(\mathbf{i}_r, \mathbf{i}_k),$$

where σ and τ are terms of $\{\mathbf{i}_1, \ldots, \mathbf{i}_{r-1}\}$. We would then have

$$\tau^{-1}\sigma = (\mathbf{i}_r, \mathbf{i}_k)(\mathbf{i}_r, \mathbf{i}_j).$$

Now the left-hand side being a term of $\{\mathbf{i}_1, \ldots, \mathbf{i}_{r-1}\}$ cannot involve the letter $\mathbf{i}_r$. The right-hand side will be in contradiction with this unless $j = k$, in which case $\tau^{-1}\sigma = \epsilon$ and $\sigma = \tau$. That is, $\sigma(\mathbf{i}_r, \mathbf{i}_j)$ can only be equal to $\tau(\mathbf{i}_r, \mathbf{i}_k)$ if $j = k$ and if $\sigma = \tau$. The $r!$ terms are therefore distinct and we have

$$\{\mathbf{i}_1, \ldots, \mathbf{i}_r\} = \{\mathbf{i}_1, \ldots, \mathbf{i}_{r-1}\}(\epsilon + (\mathbf{i}_r, \mathbf{i}_1) + \ldots + (\mathbf{i}_r, \mathbf{i}_{r-1})).$$

Similarly,

$$\{\mathbf{i}_1, \ldots, \mathbf{i}_r\}' = (\epsilon - (\mathbf{i}_r, \mathbf{i}_1) - \ldots - (\mathbf{i}_r, \mathbf{i}_{r-1}))\{\mathbf{i}_1, \ldots, \mathbf{i}_{r-1}\}'.$$

The factors on the right-hand sides of these last two equations are commutative, but we have arranged them in the order in which they will later be required.

CHAPTER II

THE CALCULUS OF TABLEAUX

Argument. Corresponding to any partition α of n a certain arrangement of n spaces, which we shall call a shape, can be constructed. By arranging the n letters in all possible ways in the n spaces of a given shape α we obtain $n!$ different tableaux S_r^α. Any two tableaux S_r^α and S_s^α of the same shape α differ only in the arrangement of the n letters in the n spaces and there is therefore some permutation σ_{rs}^α which, when applied to the tableau S_s^α, reproduces the tableau S_r^α.

Associated with any tableau S_r^α certain substitutional expressions P_r^α and N_r^α are defined, and associated with any two tableaux S_r^α and S_s^α of the same shape we can construct the substitutional expression

$$E_{rs}^\alpha \equiv P_r^\alpha \sigma_{rs}^\alpha N_s^\alpha.$$

These expressions E_{rs}^α satisfy certain important relations which are displayed in theorem 11. Included amongst these is Young's formula, which can be expressed in the form

$$P_r^\alpha N_r^\alpha P_r^\alpha N_r^\alpha = \theta^\alpha P_r^\alpha N_r^\alpha,$$

where θ^α is a certain numerical quantity which will later be evaluated.

§ 8. Shapes and Tableaux

We shall denote by $[\alpha_1, \ldots, \alpha_k]$ the *partition*

$$\alpha_1 + \ldots + \alpha_k = n, \quad \alpha_1 \geqslant \ldots \geqslant \alpha_k$$

of the number n, where $\alpha_1, \ldots, \alpha_k$ are positive integers. Thus $[4, 3, 3, 2]$ is a partition of 12. It may also be written $[4, 3^2, 2]$ and a similar notation will be used in other cases. When no confusion is likely to arise we denote the partition $[\alpha_1, \ldots, \alpha_k]$ simply by α.

Corresponding to each partition α of n we can construct a *shape*, which we may also denote by α, having α_1 spaces in the first row, α_2 in the second row and so on. Thus $[3, 2^2, 1]$ denotes the shape

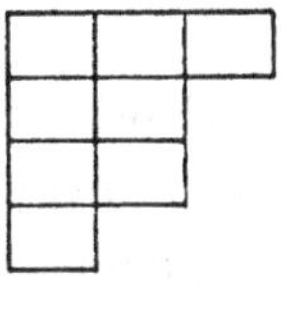

The n letters $\mathbf{1}, \ldots, \mathbf{n}$ may be arranged in the spaces of any shape α in $n!$ ways. Each such arrangement is called a *tableau*. Thus two tableaux of the shape $[3, 2^2, 1]$ are

1	2	3
4	5	
6	7	
8		

and

1	5	8
6	7	
3	4	
2		

To enable us to distinguish between different tableaux of the same shape α we arrange them in some order and call them $S_1^\alpha, \ldots, S_{n!}^\alpha$ respectively. The order chosen is at this stage immaterial, but in the next chapter when we come to deal with what are known as standard tableaux a definite sequence will be specified.

An awkward feature of our subject is the number of suffixes of one sort or another which occur. We shall adopt the plan that in any argument or formula where only one shape α is under consideration, we shall drop the suffix α unless ambiguity is caused thereby. Similarly, if we are dealing with a single tableau S_r^α we may drop both the α and the r.

Any permutation applied to a tableau of shape α will produce another tableau of the same shape. Also, any tableau can be turned into any other tableau of the same shape by some permutation. We shall denote by σ_{rs}^α the permutation which changes S_s^α into S_r^α. The reason for arranging the suffixes r, s as we have done will appear as the theory develops. Since $\sigma_{rs}{}^{-1}$ must be that permutation which changes S_r^α into S_s^α, we have the important result

$$\sigma_{rs}{}^{-1} = \sigma_{sr}.$$

Thus, if X_s^α is any substitutional expression derived from S_s^α, and X_r^α is the corresponding expression derived from S_r^α, then from theorem 2 (§ 2)

$$\sigma_{rs}^\alpha X_s^\alpha \sigma_{sr}^\alpha = X_r^\alpha.$$

It should be observed that for any given α and r (or alternatively for any given α and s), any given permutation can be expressed in the form σ_{rs}^α. Further,

$$\sigma_{rs}^\alpha \sigma_{st}^\alpha = \sigma_{rt}^\alpha,$$

that is, the change from S_t^α to S_s^α followed by that from S_s^α to S_r^α is equivalent to the change from S_t^α to S_r^α.

The arrangement of the suffixes in the last three formulae should be carefully observed as these formulae will be used many times in our subsequent theory.

§ 9. The Substitutional Expressions P and N

We shall denote by $\mathcal{P}$ the product of the positive symmetric groups of the rows of a tableau S and by $\mathcal{N}$ the product of the negative symmetric groups of the columns of S. $\mathcal{P}$ and $\mathcal{N}$ are therefore representations of certain subgroups of $\mathcal{S}_n$. The sum of all the elements of $\mathcal{P}$ will be denoted* by P and the sum of all the elements of $\mathcal{N}$ by N. Thus, if

$$S = \begin{array}{|c|c|c|} \hline 1 & 2 & 3 \\ \hline 4 & 6 & \multicolumn{1}{c}{} \\ \cline{1-2} 5 & 7 & \multicolumn{1}{c}{} \\ \cline{1-2} \end{array}\ ,$$

then

$$P = \{1, 2, 3\}\,\{4, 6\}\,\{5, 7\}, \quad N = \{1, 4, 5\}'\{2, 6, 7\}'\{3\}'.$$

In our illustration P contains 3! 2! 2! terms but N has 3! 3! 1! terms. Since no two row factors of P can involve the same letter, it follows that these factors commute and that their order is immaterial. The same applies to the column factors of N. On the other hand, a row factor of P does not necessarily commute with a column factor of N since both may involve the same letter.

We shall often find it useful to denote an element of $\mathcal{P}$ by π and one of $\mathcal{N}$ by ν. Thus any π is a permutation of $\mathcal{S}_n$ and it appears in P with the coefficient $+1$. From the manner of constructing N it will be seen that any ν must be of the form $\zeta_\tau \tau$, where τ is a permutation of $\mathcal{S}_n$. Thus ν appears in N with the coefficient $+1$ but τ appears in N with the coefficient ζ_τ. It will sometimes be advantageous to write ζ_ν in place of ζ_τ. A further simplification in notation can be effected by writing $\zeta\nu$ in place of $\zeta_\nu \nu$ in any term in which only one ν appears. In such cases it must be clearly understood that the coefficient ζ is determined by the ν factor only. With the above notation, it will be seen that we may write

$$P = \Sigma\pi, \quad N = \Sigma\nu.$$

From the fact that $\mathcal{P}$ and $\mathcal{N}$ are groups it follows that

$$\pi P = P\pi = P, \quad \nu N = N\nu = N. \qquad \ldots(9.1)$$

Again, if $\mathcal{P}$ (or $\mathcal{N}$) contains an element π (or ν), it follows from the elementary properties of groups that $\mathcal{P}$ (or $\mathcal{N}$) also contains π^{-1} (or ν^{-1}).

It will be observed † that the identity ϵ is the only permutation which can appear simultaneously in P and in N. Further, no permutation of $\mathcal{S}_n$ can arise in more than one way in the product PN, for if

$$\pi_1\nu_1 = \pi_2\nu_2,$$

then

$$\pi_2^{-1}\pi_1 = \nu_2\nu_1^{-1}.$$

But this is a permutation belonging both to P and to N and must therefore be the identity ϵ. Thus

$$\pi_2 = \pi_1, \quad \nu_2 = \nu_1.$$

* Young I, p. 133.

† Young II, p. 368.

Now P and N both have a term ϵ, so that it follows from the preceding that the coefficient of ϵ in PN is $+1$. From this we gather that the product PN does not vanish identically. The same will be true of the product NP. Since the coefficient of every permutation which occurs in P is $+1$ and that of every permutation which occurs in N is ± 1, we conclude that the coefficient of any permutation in the product PN is $+1$, -1 or 0, only.

§ 10. The Product $P_r N_s$

Consider now the expression $P_r N_s$, where $r \neq s$. If two letters **h**, **k** appear in the same row of S_r and also in the same column of S_s, then, as in theorem 4 (§ 7), P_r must have the factor $\epsilon + (\mathbf{h}, \mathbf{k})$ and N_s the factor $\epsilon - (\mathbf{h}, \mathbf{k})$. In this case

$$P_r N_s = 0$$

on account of the vanishing factor $\epsilon - (\mathbf{h}, \mathbf{k})^2$. Similarly,

$$N_s P_r = 0.$$

If no two letters occurring in the same row of S_r occur in the same column of S_s, then we shall show that it is possible to construct a tableau S_t such that

$$P_r = P_t, \quad N_s = N_t,$$

from which it follows that

$$P_r N_s = P_t N_t \neq 0.$$

In consequence of this, $P_r N_s = 0$ is not only a necessary but also a sufficient condition that P_r should have a factor $\epsilon + \tau$ and N_s a factor $\epsilon - \tau$, where τ is a transposition.

The construction of S_t is quite simple. If a letter **i** appears in the lth row of S_r and in the mth column of S_s, then **i** appears in the lth row and mth column of S_t. A mechanical method of writing down S_t is illustrated in the following example:

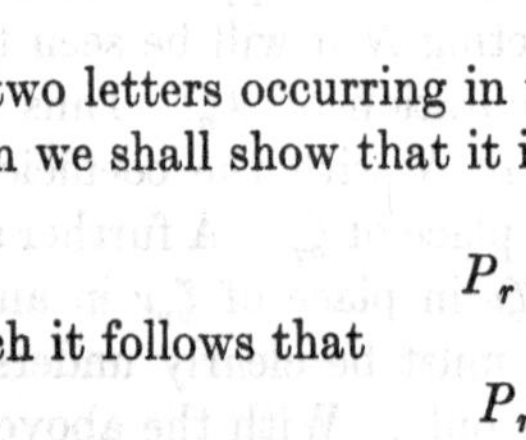

			6	5	1
			3	4	
			2	7	
1	2	4	2	4	1
3	5		3	5	
7	6		6	7	

Here the entries in S_r and S_s are given. Those of S_t are filled in one by one.

This method shows at once whether $P_r N_s$ vanishes, for in such a case it will be impossible to construct S_t. This is because the two letters **h** and **k** which occur in the lth row of S_r and in the mth column of S_s would both be forced into the same position in S_t.

Suppose again that $P_rN_s \neq 0$ and that the tableau S_t has been constructed such that $P_r = P_t$, $N_s = N_t$. It follows that σ_{tr} is an element π_r of P_r and that $\sigma_{ts} = \zeta\nu_s$ where ν_s is an element of N_s. Hence

$$\sigma_{rs} = \sigma_{rt}\sigma_{ts} = \zeta\pi_r^{-1}\nu_s, \quad \sigma_{sr} = \sigma_{st}\sigma_{tr} = \zeta\nu_s^{-1}\pi_r.$$

Now, by (9.1),

$$\pi_r^{-1}N_s\pi_r = \pi_r^{-1}\nu_s N_s \nu_s^{-1}\pi_r = \sigma_{rs}N_s\sigma_{sr} = N_r;$$

hence $\pi_r^{-1}\nu_s^{-1}\pi_r$ is in fact an element ν_r^{-1} of N_r, from which it follows at once that

$$\sigma_{sr} = \zeta\nu_s^{-1}\pi_r = \zeta\pi_r\nu_r^{-1}.$$

Similarly,

$$\sigma_{sr} = \zeta\pi_s\nu_s^{-1}.$$

In case the reader feels that there is some confusion about the meaning of ζ in these equations, we may point out that, since $\nu_r^{-1} = \pi_r^{-1}\nu_s^{-1}\pi_r$, the elements ν_r^{-1} and ν_s^{-1} belong to the same class. It follows from this that ζ has the same value whether it is determined by ν_r^{-1} or ν_s^{-1}.

Conversely, if we assume that $\sigma_{rs} = \zeta\pi_r^{-1}\nu_s$, it follows from the fact that $P_rN_r \neq 0$, that

$$P_rN_s = P_r\pi_r^{-1}\nu_s N_s = \zeta P_r\sigma_{rs}N_s = \zeta P_rN_r\sigma_{rs} \neq 0.$$

The same result can be obtained if we assume that $\sigma_{rs} = \zeta\nu_r\pi_r^{-1}$ or that $\sigma_{rs} = \zeta\nu_s\pi_s^{-1}$. The proofs in these cases are left to the reader. We have therefore now established the following important result.

THEOREM 5. *If $P_rN_s \neq 0$ (or if $N_sP_r \neq 0$), then*

$$\sigma_{sr} = \zeta\pi_s\nu_s^{-1} = \zeta\nu_s^{-1}\pi_r = \zeta\pi_r\nu_r^{-1},$$

$$\sigma_{rs} = \zeta\nu_r\pi_r^{-1} = \zeta\pi_r^{-1}\nu_s = \zeta\nu_s\pi_s^{-1},$$

and conversely.

The formulae of this theorem may be looked upon as the precise analytical equivalents of the diagrammatic process already described for the construction of S_t.

From theorem 5 we deduce without difficulty the complementary result.

THEOREM 6. *If $P_rN_s = 0$ (or if $N_sP_r = 0$), then σ_{rs} cannot be expressed in any of the forms*

$$\zeta\nu_r\pi_r^{-1}, \quad \zeta\pi_r^{-1}\nu_s, \quad \zeta\nu_s\pi_s^{-1},$$

and conversely.

§ 11. The Expressions E^{α}_{rs}

We shall find that the substitutional expressions E^{α}_{rs}, which we now define by the relation

$$E_{rs}^{\alpha} \equiv \sigma_{rs}^{\alpha} P_s^{\alpha} N_s^{\alpha} = P_r^{\alpha} \sigma_{rs}^{\alpha} N_s^{\alpha} = P_r^{\alpha} N_r^{\alpha} \sigma_{rs}^{\alpha},$$

play an important part in the subsequent theory. Evidently E_{rr}^{α} is just the product $P_r^{\alpha} N_r^{\alpha}$; and as such products appear very frequently in the following pages, we may effect a further economy in our notation by writing

$$E_{rr}^{\alpha} \equiv E_r^{\alpha}.$$

Thus

$$E_{rs}^{\alpha} = \sigma_{rs}^{\alpha} E_s^{\alpha} = E_r^{\alpha} \sigma_{rs}^{\alpha}. \qquad \ldots(11.1)$$

This formula shows that the coefficient of any permutation τ in E_{rs}^{α} must be $+1$, -1 or 0, since this is so in the case of E_r^{α}.

We shall now describe a method of evaluating E_{rs}^{α} in terms of the permutations of $\mathfrak{S}_n$. If τ is of the form $\zeta \pi_r \sigma_{rs} \nu_s$, that is, if τ occurs with a non-zero coefficient in E_{rs}, then by (9.1),

$$\begin{aligned} \tau N_s \tau^{-1} P_r &= \pi_r \sigma_{rs} \nu_s N_s \nu_s^{-1} \sigma_{sr} \pi_r^{-1} P_r \\ &= \pi_r \sigma_{rs} N_s \sigma_{sr} P_r \\ &= \pi_r N_r P_r \\ &\neq 0. \end{aligned}$$

Hence, if $\tau N_s \tau^{-1} P_r = 0$, the coefficient of τ in E_{rs} must be zero. Next, suppose that $\tau N_s \tau^{-1} P_r \neq 0$. On putting $\tau = \sigma_{us}$, and therefore $\tau^{-1} = \sigma_{su}$, we have

$$\tau N_s \tau^{-1} P_r = \sigma_{us} N_s \sigma_{su} P_r = N_u P_r \neq 0.$$

It follows from theorem 5 (§ 10) that $\sigma_{ru} = \zeta \pi_r^{-1} \nu_u$. Now

$$\sigma_{rs} = \sigma_{ru} \sigma_{us} = \zeta \pi_r^{-1} \nu_u \sigma_{us} = \zeta \pi_r^{-1} \sigma_{us} \nu_s = \zeta \pi_r^{-1} \tau \nu_s.$$

Thus

$$\tau = \zeta \pi_r \sigma_{rs} \nu_s^{-1}.$$

This means that τ occurs in E_{rs} with that coefficient ζ which was originally determined by ν_u. We summarise these results in the following rules.

THEOREM 7. *If* $N_u P_r = 0$, *the coefficient of* $\tau (= \sigma_{us})$ *in* E_{rs} *is* 0. *If* $N_u P_r (= N_t P_t) \neq 0$, *the coefficient of* τ *in* E_{rs} *is* $+1$ *or* -1 *according as* σ_{ut} *is even or odd.*

The manipulation involved may be tabulated as in the following example. We proceed to evaluate E_{12}, where

$$S_1 = \begin{array}{|c|c|} \hline 1 & 2 \\ \hline 3 & \\ \cline{1-1} \end{array}, \quad S_2 = \begin{array}{|c|c|} \hline 1 & 3 \\ \hline 2 & \\ \cline{1-1} \end{array}, \quad \sigma_{12} = \sigma_{21} = (2, 3).$$

	ϵ	(1, 2)	(2, 3)	(3, 1)	(1, 2, 3)	(3, 2, 1)	τ
	1 3 2	2 3 1	1 2 3	3 1 2	2 1 3	3 2 1	S_u
$S_1 =$ 1 2 3	.	.	1 2 3	2 1 3	2 1 3	1 2 3	S_t
	.	.	ϵ	(2, 3)	ϵ	(3, 1)	σ_{ut}
	0	0	+1	−1	+1	−1	coef. of τ in E_{12}

Hence $E_{12} = (2, 3) - (3, 1) + (1, 2, 3) - (3, 2, 1)$. A further explanation of the above diagram is perhaps necessary. The second row gives the 3! tableaux S_u obtained by applying the permutations τ to S_2. The third row gives six examples of the diagram described in the previous section for the construction of S_t. The dots in the first two places indicate that no such S_t is possible and that therefore in these cases $N_u P_1 = 0$. The fourth row gives the permutation σ_{ut} which yields S_u when it is applied to S_t.

§ 12. von Neumann's Theorem

Let $X \equiv \lambda_1 \epsilon + \lambda_2 \tau_2 + \ldots + \lambda_{n!} \tau_{n!}$ be any substitutional expression with numerical coefficients $\lambda_1, \ldots, \lambda_{n!}$. Since $\tau_j, \tau_j\tau_2, \ldots, \tau_j\tau_{n!}$ are merely the permutations $\epsilon, \tau_2, \ldots, \tau_{n!}$ in some order, the expression $\tau_j X$ differs from X only in that the coefficients $\lambda_1, \ldots, \lambda_{n!}$ have been subjected to a permutation. The same is true of the expressions $X\tau_j$ and $\tau_i X \tau_j$. For certain expressions X and certain permutations τ_i, τ_j it may be that $\tau_i X \tau_j = X$. If this is so, then, since the permutations $\epsilon, \tau_2, \ldots, \tau_{n!}$ are linearly independent, we can equate the coefficients of each permutation on both sides of the equation $\tau_i X \tau_j = X$. This is the method employed in proving the theorems of this section.

THEOREM 8 (von Neumann's Theorem *). *If $X = \pi X \nu$ for all π from $\mathfrak{P}$ and all ν from $\mathfrak{N}$, then*

$$X = \lambda PN,$$

where λ is the coefficient of ϵ in X.

PROOF. Any permutation either occurs in PN or it does not. Suppose first that τ_j does occur in PN. In this case τ_j must be of the form $\zeta_j \pi_j \nu_j$. Since by hypothesis

$$X = \pi_j X \nu_j,$$

we may equate the coefficients of τ_j on both sides of the equation

$$\sum_i \lambda_i \tau_i = \sum_i \pi_j \lambda_i \tau_i \nu_j.$$

* v. d. Waerden, § 127.

Since each permutation occurs once only on the right-hand side, and the appropriate term is that for which $\tau_i = \epsilon = \tau_1$, we therefore conclude that

$$\lambda_j = \zeta_j \lambda_1,$$

and that

$$\lambda_j \tau_j = \lambda_1 \pi_j \nu_j.$$

Secondly, if τ_j does not occur in $P_r N_r$, it cannot be of the form $\zeta \pi_r \nu_r^{-1}$. Hence if we write $\tau_j = \sigma_{sr}$, then, by theorem 6 (§ 10), $P_r N_s = 0$. When this happens there must be a transposition τ_h such that P_r has the factor $\epsilon + \tau_h$ and N_s has the factor $\epsilon - \tau_h$. Evidently τ_h is an element π_r of $\mathfrak{P}_r$ and $-\tau_h$ is an element ν_s of $\mathfrak{N}_s$. It follows that

$$-\tau_j^{-1}\tau_h\tau_j = \sigma_{rs}\nu_s\sigma_{sr} = \nu_r,$$

that is, $-\tau_j^{-1}\tau_h\tau_j$ is an element of $\mathfrak{N}_r$. Hence, by hypothesis,

$$\sum_i \lambda_i \tau_i = -\sum_i \tau_h \lambda_i \tau_i \tau_j^{-1} \tau_h \tau_j.$$

Since any permutation appears once only on the right-hand side and since $\tau_h^2 = \epsilon$, the term in τ_j must be that for which $i = j$. Thus, comparing the coefficients of τ_j on both sides of this equation, we obtain

$$\lambda_j = -\lambda_j,$$

that is,

$$\lambda_j = 0$$

if τ_j does not occur in PN.

Combining our two results, we conclude that

$$X = \lambda_1 \sum_j \pi_j \nu_j,$$

the summation being over all j for which τ_j appears in PN. In other words

$$X = \lambda_1 PN.$$

This concludes the proof of the theorem.

From von Neumann's theorem we now deduce another theorem of which von Neumann's is a special case.

THEOREM 9. *If $X = \pi_r X \nu_s$ for all π_r from $\mathfrak{P}_r$ and for all ν_s from $\mathfrak{N}_s$, then*

$$X = \lambda E_{rs},$$

where λ is the coefficient of σ_{rs} in X.

PROOF. Since $X = \pi_r X \nu_s$ for all π_r and all ν_s, $X\sigma_{sr} = \pi_r X \nu_s \sigma_{sr}$ for all π_r and ν_s. It follows that $X\sigma_{sr} = \pi_r X \sigma_{sr} \nu_r$ for all π_r and all ν_r. From von Neumann's theorem we conclude that

$$X\sigma_{sr} = \lambda P_r N_r,$$

and hence that

$$X = \lambda P_r N_r \sigma_{rs} = \lambda E_{rs},$$

where λ is the coefficient of ϵ in $X\sigma_{sr}$, that is, where λ is the coefficient of σ_{rs} in X. This completes the proof of the theorem.

We conclude this section with an important application of theorem 9. Since $\pi_r P_r = P_r$ and $N_s \nu_s = N_s$, it follows that for any X whatever

$$\pi_r P_r X N_s \nu_s = P_r X N_s$$

and

$$\pi_r E_{ru} X E_{vs} \nu_s = E_{ru} X E_{vs},$$

for all π_r from P_r and all ν_s from N_s. From theorem 9 we deduce that

$$P_r X N_s = \lambda E_{rs}$$

and

$$E_{ru} X E_{vs} = \lambda E_{rs},$$

where in each case λ is the coefficient of σ_{rs} in the left-hand side of the equation.

§ 13. Young's Formula

If we take $X = \epsilon$ and put $r = u = v = s$ in the last formula of the preceding section, we obtain

$$E_r^\alpha E_r^\alpha = \theta^\alpha E_r^\alpha,$$

where θ^α is a numerical constant. By operating on both sides of this last equation with any permutation σ_{sr}, it will be seen that θ^α is the same for all tableaux of shape α; that is θ^α is the same for all values of r. The formula just quoted was first given by Young in a slightly different form; * and as it plays a crucial part in development of the theory we may appropriately name it *Young's formula*. Young's proof of this formula is very involved but it has the merit that it provides a definite numerical value for θ^α. By leaving the evaluation of θ^α till a later stage (§ 33), we can avoid the intricacies of Young's method, but before we can proceed we must show that θ^α is non-zero. It will appear later that θ^α is in fact always a positive integer.

LEMMA. *The coefficients of ϵ in the products X_1X_2 and X_2X_1 of two substitutional expressions are the same.*

PROOF. If ϵ occur in the product X_1X_2, it can only do so as a sum of terms $(\xi_1\sigma)(\xi_2\sigma^{-1})$, where ξ_1, ξ_2 are numerical, $\xi_1\sigma$ being a term of X_1 and $\xi_2\sigma^{-1}$ a term of X_2. Precisely the same terms and no others yield terms in ϵ in the product X_2X_1. That is, the coefficients of ϵ in X_1X_2 and X_2X_1 are the same.

Consider now the expression PNP. The coefficient of ϵ in PNP is the same as the coefficient of ϵ in PPN. Since $\mathcal{P}$ is a sub-group of $\mathcal{S}_n$ of order $\alpha_1!\alpha_2!\ldots\alpha_k!$, it is clear that

$$PP = \alpha_1!\alpha_2!\ldots\alpha_k!\,P.$$

Since we know that the coefficient of ϵ in PN is $+1$, we conclude that the

* Young II, p. 366.

coefficient of ϵ in PNP is $\alpha_1!\alpha_2!\ldots\alpha_k!$. It follows from this that the product PNP does not vanish.

Let $\pi_1\nu\pi_2$ be a typical term of PNP. Its inverse $\pi_2^{-1}\nu^{-1}\pi_1^{-1}$ must also appear in the product PNP. Since ζ_ν and $\zeta_{\nu^{-1}}$ are either both $+1$ or both -1, it follows that the coefficients of σ and of σ^{-1} in PNP are the same, where σ is any permutation of $\mathcal{S}_n$. It follows from this that the coefficient of ϵ in $PNPPNP$ is the sum of squares. Since PNP does not vanish, we conclude that the coefficient of ϵ in $PNPPNP$ is non-zero and positive. Now

$$PNPPNP=\alpha_1!\alpha_2!\ldots\alpha_k!\,PNPNP.$$

Since we have shown that the coefficient of ϵ in this expression does not vanish, it follows that $PNPN$ cannot vanish. That is, $\theta^\alpha \neq 0$.

We shall conclude this section by deducing a result which will be used in the next chapter. In § 12 it was shown that

$$E_{ru}XE_{vs}=\lambda E_{rs},$$

where λ is the coefficient of σ_{rs} on the left-hand side. Thus, recalling (11.1),

$$\begin{aligned}\lambda &= \text{coefficient of } \sigma_{rs} \text{ in } \sigma_{ru}E_uXE_v\sigma_{vs}\\ &= \ldots\ldots\ldots\ldots\ \epsilon\ \ldots\ \sigma_{sr}\sigma_{ru}E_uXE_v\sigma_{vs}.\end{aligned}$$

From the lemma it follows that

$$\begin{aligned}\lambda &= \text{coefficient of } \epsilon \text{ in } E_v\sigma_{vs}\sigma_{sr}\sigma_{ru}E_uX\\ &= \ldots\ldots\ldots\ldots\ \epsilon\ \ldots\ E_v\sigma_{vu}E_uX\\ &= \ldots\ldots\ldots\ldots\ \epsilon\ \ldots\ E_vE_v\sigma_{vu}X\\ &= \ldots\ldots\ldots\ldots\ \epsilon\ \ldots\ \theta E_v\sigma_{vu}X\\ &= \ldots\ldots\ldots\ldots\ \epsilon\ \ldots\ \theta E_{vu}X.\end{aligned}$$

We have therefore demonstrated the truth of the following theorem.

THEOREM 10a. *For all values of r, u, v, s and for any substitutional expression X whatever,*

$$E^\alpha_{ru}XE^\alpha_{vs}=\theta^\alpha\rho E^\alpha_{rs},$$

where ρ is the coefficient of ϵ in $E^\alpha_{vu}X$.

In particular, as a special case of this theorem,

$$E_rXE_r=\theta\rho E_r,$$

where ρ is the coefficient of ϵ in E_rX.

§ 14. Tableaux of Different Shapes

Using the notation of § 8, let α and β be two different partitions of n. If the first of the numbers $\alpha_1-\beta_1$, $\alpha_2-\beta_2$, ... which is non-zero is positive, then we say that the partition α comes after β and we write this $\alpha>\beta$. Since we are supposing that $\alpha \neq \beta$, it follows that S_r^α and S_s^β are tableaux of different shapes.

If no two letters from the first row of S_r^α appear in the same column of S_s^β, then S_s^β must have a row with at least α_1 letters, so that $\beta_1 \geqslant \alpha_1$. If $\beta_1 > \alpha_1$, then $\beta > \alpha$. Suppose then that $\beta_1 = \alpha_1$; in this case one letter of the first row of S_r^α must appear in each column of S_s^β. If no two letters from the second row of S_r^α appear in the same column of S_s^β, then S_s^β must have, in addition to the row with α_1 $(=\beta_1)$ letters, a row with at least α_2 letters, so that $\beta_2 \geqslant \alpha_2$. If $\beta_1 = \alpha_1$, $\beta_2 > \alpha_2$, then $\beta > \alpha$, but if $\beta_1 = \alpha_1$, $\beta_2 = \alpha_2$, we proceed with the third row of S_r^α, and so on. Eventually we come to the conclusion that either $\beta > \alpha$ or else there are two letters in some row of S_r^α which appear in the same column of S_s^β. This conclusion is quite independent of the arrangement of the letters in the shapes α and β, that is, it is independent of the particular values which the suffixes r and s may have.

It follows from the preceding argument that if S_r^α, S_s^β be any two tableaux of different shapes, then either (i) $\alpha > \beta$, in which case *

$$N_s^\beta P_r^\alpha = P_r^\alpha N_s^\beta = 0, \qquad \ldots(14.1)$$

because S_r^α has in some row two letters which occur in the same column of S_s^β; or else (ii) $\alpha < \beta$, in which case

$$N_r^\alpha P_s^\beta = P_s^\beta N_r^\alpha = 0$$

for a similar reason.

Suppose now that $X \equiv \lambda_1 \sigma_{1s}^\beta + \ldots + \lambda_{n!} \sigma_{n!s}^\beta$ is any substitutional expression. If $\alpha > \beta$, it follows from the foregoing that

$$P_r^\alpha X N_s^\beta = \sum_i \lambda_i P_r^\alpha N_i^\beta \sigma_{is}^\beta = 0.$$

In particular, after replacing X by $N_r^\alpha X P_s^\beta$ in the vanishing $P_r^\alpha X N_s^\beta$, we have

$$E_r^\alpha X E_s^\beta = 0.$$

Similarly, if $\alpha < \beta$,

$$N_r^\alpha X P_s^\beta = 0; \qquad \ldots(14.2)$$

and so in this case also

$$E_r^\alpha X E_s^\beta = 0.$$

If X is now replaced by $\sigma_{ru}^\alpha X \sigma_{vs}^\beta$, we find that

$$E_{ru}^\alpha X E_{vs}^\beta = E_r^\alpha \sigma_{ru}^\alpha X \sigma_{vs}^\beta E_s^\beta = 0.$$

A combination of these two cases yields the following result which is complementary to theorem 10a.

THEOREM 10b. *If $\alpha \neq \beta$, then for all values of r, u, v, s and for any substitutional expression X whatever*

$$E_{ru}^\alpha X E_{vs}^\beta = 0.$$

* Young I, p. 134.

If we now choose $X=\epsilon$ in theorems 10a and 10b and apply theorem 7 (§ 11) in the first case, we arrive at the following important conclusion.

THEOREM 11. *The expressions E^{α}_{rs} satisfy the following relations :*

(i) $E^{\alpha}_{ru}E^{\beta}_{vs}=0$ *if* $\alpha \neq \beta$;

(ii) $E^{\alpha}_{ru}E^{\alpha}_{vs}=0$ *if* $N^{\alpha}_{u}P^{\alpha}_{v}=0$;

(iii) $E^{\alpha}_{ru}E^{\alpha}_{vs}=\pm\theta^{\alpha}E^{\alpha}_{rs}$ *if* $N^{\alpha}_{u}P^{\alpha}_{v}(=N^{\alpha}_{t}P^{\alpha}_{t})\neq 0$,

according as σ_{ut} is an even or an odd permutation.

CHAPTER III

THE SEMI-NORMAL REPRESENTATION

Argument. The crucial result of this chapter is theorem 12, which states that $n!$ linearly independent substitutional expressions e^{α}_{rs} can be constructed such that any substitutional expression whatever can be expressed as a linear combination of the units e^{α}_{rs} and such that the relations

$$e^{\alpha}_{rs} e^{\beta}_{uv} = \delta^{\alpha\beta} \delta_{su} e^{\alpha}_{rv},$$
$$\sum_{\alpha, r} e^{\alpha}_{rr} = \epsilon,$$

which employ the Kronecker delta, are satisfied. As is indicated by the notation, each e^{α}_{rs} is associated with a certain shape α and with two tableaux S^{α}_{r} and S^{α}_{s} of this shape. Now in Chapter II we defined $n!$ different tableaux of each shape, so that without further qualification we should have far more tableaux than are required. The concept of a standard tableau is accordingly introduced. Of these there are a certain number f^{α} associated with each shape α; and it is shown that

$$\sum_{\alpha} (f^{\alpha})^2 = n!.$$

When our attention is restricted to standard tableaux only, it is found that we have exactly $n!$ expressions e^{α}_{rs} as is required by theorem 12.

If we can associate with the permutations $\sigma_i, \sigma_j, \ldots$ of the symmetric group a set of matrices $U_{\sigma_i}, U_{\sigma_j}, \ldots$ such that

$$U_{\sigma_i} U_{\sigma_j} = U_{\sigma_i \sigma_j},$$

the set of matrices is said to form a representation of the group. It is shown that the relations

$$e^{\alpha}_{rs} e^{\beta}_{uv} = \delta^{\alpha\beta} \delta_{su} e^{\alpha}_{rv}$$

mentioned above lead to one such representation for every partition α and that, furthermore, each of these representations is irreducible. Such matrices, besides being of fundamental theoretical importance, have practical applications in the analysis of symmetry and in the solution of substitutional equations.

The irreducible representation associated with a particular partition α is however by no means unique, although all such representations are equivalent. Among the possible representations are the semi-normal, the orthogonal and the natural representations. The semi-normal one is that which is derived in this chapter, and the associated semi-normal units e^{α}_{rs} are given explicitly. Young derived these semi-normal units from the natural ones, but here we derive the semi-normal units directly, following to some extent the method devised by Thrall for the construction of the orthogonal representation. Should the reader desire to pursue the natural representation only, he can substitute § 28 and § 29 of Chapter IV for § 17 of the present chapter, and thereafter, in §§ 18, 19, interpret e^{α}_{rs}, $u^{\alpha}_{rs\tau_i}$ in terms of the natural representation. He can then proceed to Chapters V and VI. The remaining part of this chapter, §§ 20-23, is devoted to the proof of theorem 18 which shows how a semi-normal matrix representing a transposition of two consecutive letters takes a very simple form. Since any permutation may be expressed as a product of such transpositions, this theorem gives a method whereby we may construct the semi-normal matrix of any permutation or indeed of any substitutional expression whatever.

§ 15. Standard Tableaux

Of the $n!$ tableaux of shape α there will be a certain number f^{α} which have the property that the letters in each row and in each column are in lexical order. Such tableaux are called *standard tableaux*.* We shall find it necessary to arrange all the standard tableaux of a given shape according to what is known as the *last letter sequence*. From the definition of a standard tableau it is evident that the last letter **n** must appear at the extreme right of some row and at the foot of some column. In the last letter sequence all tableaux which have the last letter **n** in the last row come before those in which **n** appears in the second-last row. These latter come before those which have **n** in the third-last row and so on. Next, those tableaux which have **n** in the same row are arranged by the same plan according to the position of the second-last letter **n − 1**. Those which have **n** and **n − 1** in the same positions are arranged according to the position of **n − 2** and so on. Illustrations of standard tableaux arranged according to the last letter sequence will be found in § 28.

It is to be understood that all tableaux referred to in this and subsequent chapters are standard unless the contrary is stated. This being so, no confusion with the notation of the last chapter will arise if we denote the f standard tableaux of a given shape, when arranged according to the last letter sequence, by

$$S_1, S_2, \ldots, S_f.$$

Before proceeding, we shall prove certain important relations between the numbers f^{α}. If from any standard tableau of shape α we remove the last letter **n** we automatically obtain a standard tableau involving $n-1$ letters. In fact all tableaux of shape α having **n** in row i will yield new tableaux of shape $[\alpha_1, \ldots, \alpha_i - 1, \ldots, \alpha_k]$. Thus, if we denote this shape by $[\alpha_i -]$ for the sake of convenience, the preceding argument shows that †

$$f^{\alpha} = \sum_{i=1}^{k} f^{[\alpha_i -]}. \qquad \ldots (15.1)$$

* Young III, p. 258. † Young III, p. 261.

A word of explanation is necessary here, for if $\alpha_i = \alpha_{i+1}$ then $\alpha_i - 1 < \alpha_{i+1}$. Since we do not recognise tableaux in which any row is shorter than the one following it, we must define $f^{[\alpha_i -]} \equiv 0$ in such cases if we wish to sum from 1 to k in the above formula. It is of course evident that in such a case **n** cannot appear in row i.

We now prove by induction the relation *

$$(n+1)f^{\alpha} = \sum_{i=1}^{k+1} f^{[\alpha_i +]}, \qquad \ldots(15.2)$$

where $[\alpha_i +]$ denotes that partition $[\alpha_1, \ldots, \alpha_i + 1, \ldots, \alpha_{k+1}]$ of $n+1$ in which α_{k+1} vanishes and where we define $f^{[\alpha_i +]} \equiv 0$ in the case where $\alpha_i = \alpha_{i-1}$. Now from (15.1) we may write

$$f^{[\alpha_i +]} = \sum_j f^{[\alpha_i +,\, \alpha_j -]},$$

where $[\alpha_i +, \alpha_j -]$ denotes the partition $[\alpha_1, \ldots, \alpha_i + 1, \ldots, \alpha_j - 1, \ldots, \alpha_{k+1}]$ when $i \neq j$ and where $[\alpha_i +, \alpha_i -]$ is the partition α itself. In the last formula j is summed from 1 to k except when $i = k+1$, in which case j is summed from 1 to $k+1$. Also by the induction hypothesis we have

$$nf^{[\alpha_j -]} = \sum_{i=1}^{k+1} f^{[\alpha_i +, \alpha_j -]}.$$

A combination of the last two formulae now gives

$$\begin{aligned} \sum_{i=1}^{k+1} f^{[\alpha_i +]} &= \sum_{i=1}^{k+1} \sum_{j=1}^{k} f^{[\alpha_i +, \alpha_j -]} + f^{[\alpha_{k+1} +, \alpha_{k+1} -]} \\ &= \sum_{j=1}^{k} nf^{[\alpha_j -]} + f^{\alpha} \\ &= (n+1)f^{\alpha}, \end{aligned}$$

which establishes the induction. Since (15.2) may be readily verified when $n = 1, 2$, the formula (15.2) holds in general.

From the results just proved it is evident that

$$(n+1)\sum_{\alpha} (f^{\alpha})^2 = \sum_{\alpha} \sum_{i=1}^{k+1} f^{\alpha} f^{[\alpha_i +]}$$

and that

$$\sum_{\beta} (f^{\beta})^2 = \sum_{\beta} \sum_{i=1}^{l} f^{[\beta_i -]} f^{\beta},$$

where l has the same significance with regard to β as k has with regard to α. Now a moment's consideration will show that, if β ranges over partitions of $n+1$ while α ranges over partitions of n, the right-hand sides of these two

* Young III, p. 262.

equations are identical. Equating the left-hand sides of these equations, an induction is established which yields the formula

$$\sum_{\alpha} (f^{\alpha})^2 = n!.$$

Since this formula can be verified when $n=1, 2$, it must hold in general.

§ 16. The Evaluation of f^{α}

The actual evaluation of f^{α} is of minor importance in our theory but is of value in practical applications. We shall therefore establish the result *

$$f^{\alpha} = \frac{n! \prod_{r<s\leq k} (x_r - x_s)}{\prod_r x_r!}, \qquad \ldots (16.1)$$

where we write

$$x_r \equiv \alpha_r + k - r.$$

The proof which follows is a modification of the one used by Young and is due to Etherington and Ledermann.† The proof is again by induction and the reader may verify the truth of the formula when $n=1, 2$. It will be observed that in the case where $\alpha_i = \alpha_{i+1}$ the formula (16.1) yields $f^{[\alpha_i -]} = 0$ on account of the vanishing factor $(x_i - x_{i+1})$. This is in agreement with our previous definition. We may mention here also that if we write $[\alpha_1, \ldots, \alpha_k]$ in the form $[\alpha_1, \ldots, \alpha_k, \alpha_{k+1}]$ with α_{k+1} equal to zero, the right-hand side of (16.1) retains the same numerical value.

Assuming the truth of (16.1) in the case of $n-1$ letters, we can readily show that

$$f^{[\alpha_i -]} = \psi \frac{x_i}{n} \prod_{s \neq i} \frac{(x_i - 1 - x_s)}{(x_i - x_s)},$$

where ψ denotes the right-hand side of (16.1). We have to show that ψ is in fact the number f^{α} of standard tableaux associated with the partition α of n. From (15.1) we deduce that

$$f^{\alpha} = \frac{\psi}{n} \sum_i x_i \prod_{s \neq i} \frac{(x_i - 1 - x_s)}{(x_i - x_s)} = -\frac{\psi}{n} \sum_i \frac{x_i \phi(x_i - 1)}{\phi'(x_i)}, \qquad \ldots (16.2)$$

where we write

$$\phi(x) \equiv (x - x_1) \ldots (x - x_k).$$

Now since $x^2\phi(x-1)$ is of degree $k+2$ in x, we can write

$$\frac{x^2\phi(x-1)}{\phi(x)} = Q(x) + \frac{R(x)}{\phi(x)}, \qquad \ldots (16.3)$$

where the quotient $Q(x)$ is of degree 2 and the remainder $R(x)$ is of degree

* Young III, p. 260. † Rutherford I, p. 52.

$k-1$ or less. On multiplying throughout by $\phi(x)$ and then putting $x=x_i$, we have immediately

$$R(x_i)=x_i^2\phi(x_i-1).$$

Since the zeros $x_1, \ldots, x_k$ of $\phi(x)$ are distinct, it follows from the theory of partial fractions that

$$\frac{R(x)}{\phi(x)}=\sum_{i=1}^{k}\frac{R(x_i)}{(x-x_i)\phi'(x_i)}=\sum_{i=1}^{k}\frac{x_i^2\phi(x_i-1)}{(x-x_i)\phi'(x_i)}.$$

Substituting this value in (16.3) and putting $x=0$, we have

$$\sum_i \frac{x_i\phi(x_i-1)}{\phi'(x_i)}=Q(0). \qquad \ldots(16.4)$$

Now for suitable values of x,

$$\frac{\phi(x-1)}{\phi(x)}=\prod_i\left(1-\frac{1}{x-x_i}\right)=\prod_i\left(1-\frac{1}{x}-\frac{x_i}{x^2}-\ldots\right),$$

and $Q(0)$ is the coefficient of x^{-2} in this product. Thus

$$Q(0)=\tfrac{1}{2}k(k-1)-\sum_i x_i=-n, \qquad \ldots(16.5)$$

since

$$\Sigma x_i=\sum_i(\alpha_i+k-i)=n+\tfrac{1}{2}k(k-1).$$

Combining (16.2), (16.4) and (16.5), we obtain the desired result,

$$f^\alpha=\psi.$$

§ 17. The Semi-normal Units e^α_{rs}

As we have already pointed out in § 15, the removal of the last letter **n** from a standard tableau S yields a standard tableau involving $n-1$ letters. For our present purposes we shall denote this new tableau by S^*. Similarly, removing the letter **n** $-$ **1** from S^* we obtain S^{**} and so on, thus acquiring a sequence of standard tableaux S, S^*, $S^{**}, \ldots, S^{[1]}$ involving respectively n, $n-1$, $n-2, \ldots, 1$ letters. In terms of these tableaux we define a corresponding sequence of substitutional expressions e, e^*, $e^{**}, \ldots, e^{[1]}$ by the relations †

$$\left.\begin{aligned} e &\equiv (1/\theta)e^*Ee^*,\\ e^* &\equiv (1/\theta^*)e^{**}E^*e^{**},\\ &\cdots\cdots\cdots\cdots\cdots\\ e^{[1]} &\equiv \epsilon, \end{aligned}\right\} \qquad \ldots(17.1)$$

where, with an obvious notation, $E=PN$, $E^*=P^*N^*, \ldots$ In this way we can define an element e^α_r associated with each standard tableau S^α_r involving n letters. The following illustration should make this clear.

† Thrall, p. 614.

$$S=\begin{array}{|c|c|}\hline 1 & 3 \\ \hline 2 & \\ \cline{1-1}\end{array}, \quad S^*=\begin{array}{|c|}\hline 1 \\ \hline 2 \\ \hline\end{array}, \quad S^{**}=S^{[1]}=\begin{array}{|c|}\hline 1 \\ \hline\end{array}\ ;$$

$$E=\{\mathbf{1},\mathbf{3}\}\{\mathbf{1},\mathbf{2}\}', \quad E^*=\{\mathbf{1},\mathbf{2}\}'\ ;$$

$$\theta=3, \qquad \theta^*=2\ ;$$

$$e=\tfrac{1}{12}\{\mathbf{1},\mathbf{2}\}'\{\mathbf{1},\mathbf{3}\}\{\mathbf{1},\mathbf{2}\}'\{\mathbf{1},\mathbf{2}\}', \quad e^*=\tfrac{1}{2}\{\mathbf{1},\mathbf{2}\}', \quad e^{**}=e^{[1]}=\epsilon.$$

It should be observed here that in general $\sigma_{rs}e_s\sigma_{sr}\neq e_r$. This is because e_r is not defined by S_r alone but by the set $S_r, S_r^*, \ldots$

Our first task is to show that these expressions e_r^α satisfy the relations †

$$e_r^\alpha e_s^\beta=\delta^{\alpha\beta}\delta_{rs}e_r^\alpha, \qquad \ldots(17.2)$$

$$\sum_{\alpha,r} e_r^\alpha=\epsilon, \qquad \ldots(17.3)$$

where we make use of the Kronecker delta. That is to say, in the terminology of abstract algebra, the expressions e_r^α form a set of normal idempotents. In point of fact they are also primitive idempotents but this does not appear at the present stage.

Suppose that the row in S which contains the letter $\mathbf{n}$ consists of the letters $\mathbf{a}_1, \ldots, \mathbf{a}_j, \mathbf{n}$ and that the column in S which contains the letter $\mathbf{n}$ consists of the letters $\mathbf{b}_1, \ldots, \mathbf{b}_k, \mathbf{n}$. The expressions P and P^* differ only in that P has a factor $\{\mathbf{a}_1, \ldots, \mathbf{a}_j, \mathbf{n}\}$ where P^* has the factor $\{\mathbf{a}_1, \ldots, \mathbf{a}_j\}$. Applying the result obtained at the end of § 7, we see that

$$P=P^*(\epsilon+(\mathbf{n},\mathbf{a}_1)+\ldots+(\mathbf{n},\mathbf{a}_j)).$$

In the same way it can be shown that

$$N=(\epsilon-(\mathbf{n},\mathbf{b}_1)-\ldots-(\mathbf{n},\mathbf{b}_k))N^*.$$

Hence E is expressible as

$$\begin{aligned} PN &= P^*[\epsilon+\sum_r(\mathbf{n},\mathbf{a}_r)-\sum_s(\mathbf{n},\mathbf{b}_s)-\sum_{r,s}(\mathbf{a}_r,\mathbf{n},\mathbf{b}_s)]N^* \\ &= P^*N^*+\text{terms involving the letter } \mathbf{n}. \end{aligned}$$

Since e^* does not involve the letter $\mathbf{n}$, we deduce from the foregoing that the coefficients of ϵ in Ee^* and in E^*e^* are the same. We therefore conclude from theorem 10a (§ 13) that

$$Ee^*E=\theta\rho E, \quad E^*e^*E^*=\theta^*\rho E^*,$$

where ρ has the same numerical value in both equations.

We can further prove by induction that $\rho=1$; for if we assume that $\rho^*=1$, that is, $E^*e^{**}E^*=\theta^*E^*$, then

$$\theta^*\rho E^*=E^*e^*E^*=(1/\theta^*)E^*e^{**}E^*e^{**}E^*=E^*e^{**}E^*=\theta^*E^*.$$

† Thrall, p. 614.

This yields the required result. The induction has a basis since

$$E^{[1]}e^{[1]}E^{[1]} = \epsilon\epsilon\epsilon = \epsilon = E^{[1]} = \theta^{[1]}E^{[1]}.$$

We have therefore established the formulae

$$Ee^*E = \theta E, \quad E^*e^*E^* = \theta^*E^*.$$

Next, we prove by induction that e is idempotent, that is $e^2 = e$. The basis in this case is given by

$$(e^{[1]})^2 = \epsilon^2 = \epsilon = e^{[1]};$$

we therefore assume that $(e^*)^2 = e^*$. It follows that

$$e^2 = (1/\theta^2)e^*Ee^*Ee^* = (1/\theta)e^*Ee^* = e.$$

From the foregoing, two other useful results may be derived, namely,

$$EeE = (1/\theta)Ee^*Ee^*E = Ee^*E = \theta E,$$
$$eEe = (1/\theta^2)e^*Ee^*Ee^*Ee^* = e^*Ee^* = \theta e.$$

We now define new elements e^α_{rs} by the equation

$$e^\alpha_{rs} \equiv (1/\theta^\alpha)e^{\alpha*}_r E^\alpha_{rs} e^{\alpha*}_s,$$

where e^α_{rr} is the idempotent previously denoted by e^α_r. The relation (17.2) is included in the formula

$$e^\alpha_{rs} e^\beta_{uv} = \delta^{\alpha\beta}\delta_{su} e^\alpha_{rv}, \qquad \ldots(17.4)$$

which we shall now prove. This formula can be verified by direct calculation when $n = 2$ and we proceed as usual by induction, assuming that the formula holds for the case of $n-1$ letters. To clarify the following proof it should be pointed out that if the tableaux $S^{\alpha*}_s$ and $S^{\beta*}_u$ are identical, then either S^α_s and S^β_u are identical or else they are obtained from $S^{\alpha*}_s$ by adding the letter **n** to different rows, in which case $\alpha \neq \beta$. Thus if $\alpha = \beta$, $s \neq u$ we can conclude from the induction hypothesis that

$$e^{\alpha*}_s e^{\beta*}_u = 0.$$

Since

$$e^\alpha_{rs} e^\beta_{uv} = (1/\theta^\alpha)(1/\theta^\beta) e^{\alpha*}_r E^\alpha_{rs} e^{\alpha*}_s e^{\beta*}_u E^\beta_{uv} e^{\beta*}_v,$$

it follows from theorem 10b (§ 14) that $e^\alpha_{rs}e^\beta_{uv} = 0$ if $\alpha \neq \beta$, while if $\alpha = \beta$, $s \neq u$ the same is true on account of the induction hypothesis. In the remaining case $\alpha = \beta$, $s = u$ and we have

$$\begin{aligned} e^\alpha_{rs} e^\alpha_{sv} &= (1/\theta^\alpha)^2 e^{\alpha*}_r \sigma^\alpha_{rs} E^\alpha_s e^{\alpha*}_s E^\alpha_s \sigma^\alpha_{sv} e^{\alpha*}_v \\ &= (1/\theta^\alpha) e^{\alpha*}_r \sigma^\alpha_{rs} E^\alpha_s \sigma^\alpha_{sv} e^{\alpha*}_v \\ &= (1/\theta^\alpha) e^{\alpha*}_r E^\alpha_{rv} e^{\alpha*}_v \\ &= e^\alpha_{rv}. \end{aligned}$$

The formula (17.4) has now been established in all cases.

The next two formulae are generalisations of the relations $EeE = \theta E$ and $eEe = \theta e$ previously obtained. First,

$$\begin{aligned} E_r e_{rs} E_s &= (1/\theta) E_r e_r{}^* E_{rs} e_s{}^* E_s \\ &= (1/\theta) E_r e_r{}^* E_r \sigma_{rs} e_s{}^* E_s \\ &= E_r \sigma_{rs} e_s{}^* E_s \\ &= \sigma_{rs} E_s e_s{}^* E_s \\ &= \theta \sigma_{rs} E_s \\ &= \theta E_{rs}. \end{aligned}$$

Also,

$$\begin{aligned} e_r E_{rs} e_s &= (1/\theta) e_r{}^* E_r e_r{}^* E_r \sigma_{rs} e_s \\ &= e_r{}^* E_r \sigma_{rs} e_s \\ &= e_r{}^* \sigma_{rs} E_s e_s \\ &= (1/\theta) e_r{}^* \sigma_{rs} E_s e_s{}^* E_s e_s{}^* \\ &= e_r{}^* \sigma_{rs} E_s e_s{}^* \\ &= e_r{}^* E_{rs} e_s{}^* \\ &= \theta e_{rs}. \end{aligned} \qquad \ldots(17.5)$$

By combining these last two results with the lemma of § 13 it is now possible to determine the coefficient of ϵ in e^{α}_{rs}. In fact, since (§ 9) the coefficient of ϵ in E_r is $+1$, we have

$$\begin{aligned} &\text{the coefficient of } \epsilon \text{ in } e_{rs} \\ &= \ldots\ldots\ldots\ldots (1/\theta) e_r E_{rs} e_s \\ &= \ldots\ldots\ldots\ldots (1/\theta) e_s e_r E_{rs} \\ &= \ldots\ldots\ldots\ldots (1/\theta) \delta_{rs} e_r E_{rs} \\ &= \ldots\ldots\ldots\ldots (1/\theta) \delta_{rs} e_r E_r \\ &= \ldots\ldots\ldots\ldots (1/\theta^2) \delta_{rs} e_r E_r E_r \\ &= \ldots\ldots\ldots\ldots (1/\theta^2) \delta_{rs} E_r e_r E_r \\ &= \ldots\ldots\ldots\ldots (1/\theta) \delta_{rs} E_r \\ &= \delta_{rs} (1/\theta). \end{aligned} \qquad \ldots(17.6)$$

We now have in all $\sum_{\alpha}(f^{\alpha})^2$ or $n!$ different expressions e^{α}_{rs}. None of them vanish since, for all values of r and s, $E^{\alpha}_r e^{\alpha}_{rs} E^{\alpha}_s = \theta^{\alpha} E^{\alpha}_{rs} \neq 0$. These expressions are linearly independent: for if $\sum_{\alpha, r, s} \lambda^{\alpha}_{rs} e^{\alpha}_{rs} = 0$, then

$$e^{\alpha}_u \sum_{\alpha, r, s} \lambda^{\alpha}_{rs} e^{\alpha}_{rs} e^{\beta}_v = \lambda^{\beta}_{uv} e^{\beta}_{uv} = 0,$$

from which it follows that all the coefficients λ^{α}_{rs} vanish, and that consequently no such linear relation can exist. We conclude, therefore, that every permutation of $\mathcal{S}_n$, and therefore any substitutional expression whatever, can be expressed as a linear combination of the e^{α}_{rs}. We return to this important fact in a moment, but for the present we remark that in particular

$$e^* = \sum_{\alpha, r, s} \lambda^{\alpha}_{rs} e^{\alpha}_{rs},$$

where the coefficients λ^{α}_{rs} are still to be determined. It follows from (17.4) that

$$e^{\alpha}_{r} e^* e^{\alpha}_{s} = \lambda^{\alpha}_{rs} e^{\alpha}_{rs}.$$

Now if either $S^{\alpha *}_{r}$ or $S^{\alpha *}_{s}$ is different from S^*, then from (17.1) and (17.2) we may show that $\lambda^{\alpha}_{rs} = 0$. That is, λ^{α}_{rs} vanishes unless $S^{\alpha}_{r} = S^{\alpha}_{s}$ and $S^{\alpha *}_{r} = S^*$. In this remaining case

$$\begin{aligned} e^{\alpha}_{r} e^* e^{\alpha}_{r} &= (1/\theta^{\alpha}) e^* E^{\alpha}_{r} e^* e^* e^{\alpha}_{r} \\ &= (1/\theta^{\alpha}) e^* E^{\alpha}_{r} e^* e^{\alpha}_{r} \\ &= e^{\alpha}_{r} e^{\alpha}_{r} \\ &= e^{\alpha}_{r}, \end{aligned} \qquad \ldots(17.7)$$

and so $\lambda^{\alpha}_{rr} = 1$. Thus $e^* = \Sigma e^{\alpha}_{r}$ where the summation ranges over those tableaux S^{α}_{r} only which are obtained from S^* by the addition of the letter **n** at the right of some row and at the bottom of some column.† Since each tableau S^{α}_{r} of n letters can be obtained in this manner in one way only from a tableau S^{β}_{s} of $n-1$ letters, it follows that

$$\sum_{\alpha, r} e^{\alpha}_{r} = \sum_{\beta, s} e^{\beta}_{s},$$

and hence, by induction, that

$$\sum_{\alpha, r} e^{\alpha}_{r} = \sum_{\beta, s} e^{\beta}_{s} = \ldots = e^{[1]}_{1} = \epsilon.$$

This establishes the formula (17.3). To clarify the import of this result we may consider the case $n=3$. Here there are four standard tableaux of three different shapes, namely,

$$S^{[3]}_{1} = \begin{array}{|c|c|c|}\hline 1 & 2 & 3 \\ \hline\end{array}, \quad S^{[2,1]}_{1} = \begin{array}{|c|c|}\hline 1 & 2 \\ \hline 3 \\ \cline{1-1}\end{array}, \quad S^{[2,1]}_{2} = \begin{array}{|c|c|}\hline 1 & 3 \\ \hline 2 \\ \cline{1-1}\end{array}, \quad S^{[1^3]}_{1} = \begin{array}{|c|}\hline 1 \\ \hline 2 \\ \hline 3 \\ \hline\end{array},$$

and two standard tableaux which involve two letters, namely,

$$S^{[2]}_{1} = \begin{array}{|c|c|}\hline 1 & 2 \\ \hline\end{array}, \quad S^{[1^2]}_{1} = \begin{array}{|c|}\hline 1 \\ \hline 2 \\ \hline\end{array}.$$

The tableaux $S^{[3]}_{1}$ and $S^{[2,1]}_{1}$ have $S^{[2]}_{1}$ embedded in them while $S^{[2,1]}_{2}$ and $S^{[1^3]}_{1}$ have $S^{[1^2]}_{1}$ embedded.

Thus

$$e^{[3]}_{1} + e^{[2,1]}_{1} = e^{[2]}_{1}, \quad e^{[2,1]}_{2} + e^{[1^3]}_{1} = e^{[1^2]}_{1}.$$

Similarly,

$$e^{[2]}_{1} + e^{[1^2]}_{1} = e^{[1]}_{1} = \epsilon.$$

Hence

$$e^{[3]}_{1} + e^{[2,1]}_{1} + e^{[2,1]}_{2} + e^{[1^3]}_{1} = e^{[2]}_{1} + e^{[1^2]}_{1} = \epsilon.$$

† Thrall, p. 615.

The expressions e^{α}_{rs} which we have introduced in this section are Young's *semi-normal units*. They are called semi-normal because the invariant quadratic (§ 25) associated with them has a matrix which is diagonal but not scalar. They are the bricks with which we shall build the theory. We therefore summarise the more important results of this section in the following theorem.

THEOREM 12. *The $n!$ semi-normal units e^{α}_{rs} are linearly independent, and any substitutional expression can be expressed as a linear combination of them. These semi-normal units satisfy the following relations :*

$$e^{\alpha}_{rs}e^{\beta}_{uv}=\delta^{\alpha\beta}\delta_{su}e^{\alpha}_{rv}, \qquad \ldots(17.4)$$

$$\sum_{\alpha,r} e^{\alpha}_{r}=\epsilon. \qquad \ldots(17.3)$$

As we have already mentioned, the expressions e^{α}_{r} form a set of *normal idempotents*, that is to say, they satisfy equation (17.2). We can now show that each e^{α}_{r} is in fact a *primitive idempotent*, that is to say, it is not possible to find expressions e' and e'', neither of which vanish, such that

$$e^{\alpha}_{r}=e'+e'', \quad e'^{2}=e', \quad e''^{2}=e'', \quad e'e''=e''e'=0. \qquad \ldots(17.8)$$

Suppose that these relations were valid ; then, writing, according to theorem 12,

$$e'=\sum_{\beta,u,v}\lambda^{\beta}_{uv}e^{\beta}_{uv},$$

where λ^{β}_{uv} is numerical, we should have, by (17.4) and (17.8),

$$e'=e^{\alpha}_{r}e'e^{\alpha}_{r}=e^{\alpha}_{r}(\sum_{\beta,u,v}\lambda^{\beta}_{uv}e^{\beta}_{uv})e^{\alpha}_{r}=\lambda^{\alpha}_{rr}e^{\alpha}_{r}$$

and

$$e'=e'^{2}=(\lambda^{\alpha}_{rr}e^{\alpha}_{r})^{2}=(\lambda^{\alpha}_{rr})^{2}e^{\alpha}_{r}.$$

Thus $(\lambda^{\alpha}_{rr})^{2}=\lambda^{\alpha}_{rr}$, and so $\lambda^{\alpha}_{rr}=1$ or 0.

If $\lambda^{\alpha}_{rr}=0$, then $e'=0$; while if $\lambda^{\alpha}_{rr}=1$, then $e'=e^{\alpha}_{r}$ and $e''=0$. Since our hypothesis was that neither e' nor e'' should vanish, we have demonstrated that e^{α}_{r} is in fact a primitive idempotent.

§ 18. Certain Fundamental Formulae

Suppose that, by resolution into its ultimate elements,

$$e^{\alpha}_{rs}=\sum_{\tau_i}v^{\alpha}_{rs\tau_i}\tau_i,$$

where the τ_i are the permutations of $\mathcal{S}_n$ and the coefficients $v^{\alpha}_{rs\tau_i}$ are numerical. According to theorem 12 every τ_i can be expressed as a linear combination of the e^{α}_{rs}. Suppose then that

$$\tau_i=\sum_{\alpha,r,s}u^{\alpha}_{rs\tau_i}e^{\alpha}_{rs}. \qquad \ldots(18.1)$$

Several important results can be obtained by combining the last two equations. It is evident that they represent reciprocal transformations.

Since $\tau_i\tau_i^{-1}=\epsilon$, we have, by using (17.3) and (17.4),

$$\sum_{\alpha,p,q,r} u^{\alpha}_{pq\tau_i} u^{\alpha}_{qr\tau_i^{-1}} e^{\alpha}_{pr} = \sum_{\alpha,r} e^{\alpha}_{rr},$$

whence

$$\sum_{q} u^{\alpha}_{pq\tau_i} u^{\alpha}_{qr\tau_i^{-1}} = \delta_{pr}.$$

Also, by expressing e^{α}_{rs} in terms of the τ_j,

$$\tau_i = \sum_{\alpha,r,s,\tau_j} u^{\alpha}_{rs\tau_i} v^{\alpha}_{rs\tau_j} \tau_j,$$

whence, since the τ_j are linearly independent,

$$\sum_{\alpha,r,s} u^{\alpha}_{rs\tau_i} v^{\alpha}_{rs\tau_j} = \delta_{\tau_i\tau_j}.$$

Again, by expressing τ_j in terms of the e^{α}_{rs},

$$e^{\alpha}_{rs} = \sum_{\beta,q,p,\tau_i} v^{\alpha}_{rs\tau_i} u^{\beta}_{qp\tau_i} e^{\beta}_{qp},$$

whence, since the e^{α}_{rs} are linearly independent,

$$\sum_{\tau_i} v^{\alpha}_{rs\tau_i} u^{\beta}_{qp\tau_i} = \delta^{\alpha\beta}\delta_{rq}\delta_{sp}. \qquad \ldots(18.2)$$

Lastly, since $e^{\alpha}_{rs}e^{\beta}_{pq}=\delta^{\alpha\beta}\delta_{sp}e^{\alpha}_{rq}$, we must have

$$\sum_{\tau_i,\tau_j} v^{\alpha}_{rs\tau_i} v^{\beta}_{pq\tau_j} \tau_i\tau_j = \delta^{\alpha\beta}\delta_{sp}\sum_{\tau_h} v^{\alpha}_{rq\tau_h}\tau_h,$$

or

$$\sum_{\tau_i,\tau_h} v^{\alpha}_{rs\tau_i} v^{\beta}_{pq(\tau_i^{-1}\tau_h)} \tau_h = \delta^{\alpha\beta}\delta_{sp}\sum_{\tau_h} v^{\alpha}_{rq\tau_h}\tau_h,$$

whence

$$\sum_{\tau_i} v^{\alpha}_{rs\tau_i} v^{\beta}_{pq(\tau_i^{-1}\tau_h)} = \delta^{\alpha\beta}\delta_{sp} v^{\alpha}_{rq\tau_h}.$$

Putting $\tau_h=\epsilon$ in this last equation and recalling from (17.6) that $v^{\alpha}_{rq\epsilon}=\delta_{rq}(1/\theta^{\alpha})$, we obtain

$$\sum_{\tau_i} v^{\alpha}_{rs\tau_i} v^{\beta}_{pq\tau_i^{-1}} = \delta^{\alpha\beta}\delta_{sp}\delta_{rq}(1/\theta^{\alpha}). \qquad \ldots(18.3)$$

Now since the permutations τ_i and the elements e^{α}_{rs} are two sets of $n!$ linearly independent elements, we can solve the equations (18.2) uniquely for the $u^{\beta}_{qp\tau_i}$ in terms of the $v^{\alpha}_{rs\tau_i}$. But this solution is exhibited in (18.3), from which it follows that *

$$u^{\beta}_{qp\tau_i} = \theta^{\beta} v^{\beta}_{pq\tau_i^{-1}}.$$

We can now discard the v-coefficients by expressing them in terms of the u-coefficients in the preceding equations. We therefore gather the following relations together :

$$e^{\alpha}_{rs} = (1/\theta^{\alpha})\sum_{\tau_i} u^{\alpha}_{sr\tau_i^{-1}}\tau_i, \qquad \ldots(18.4)$$

$$\tau_i = \sum_{\alpha,r,s} u^{\alpha}_{rs\tau_i} e^{\alpha}_{rs}, \qquad \ldots(18.1)$$

$$\sum_{q} u^{\alpha}_{pq\tau_i} u^{\alpha}_{qr\tau_i^{-1}} = \delta_{pr}, \qquad \ldots(18.5)$$

* Young IV, p. 256.

$$\sum_{\alpha, r, s} (1/\theta^{\alpha}) u^{\alpha}_{rs\tau_i} u^{\alpha}_{sr\tau_j^{-1}} = \delta_{\tau_i \tau_j}, \qquad \ldots(18.6)$$

$$\sum_{\tau_i} u^{\alpha}_{sr\tau_i^{-1}} u^{\beta}_{qp\tau_i} = \delta^{\alpha\beta} \delta_{sp} \delta_{rq} \theta^{\alpha}, \qquad \ldots(18.7)$$

$$\sum_{\tau_i} u^{\alpha}_{sr\tau_i^{-1}} u^{\beta}_{qp\tau_h \tau_i} = \delta^{\alpha\beta} \delta_{sp} \theta^{\alpha} u^{\alpha}_{qr\tau_h}, \qquad \ldots(18.8)$$

$$u^{\alpha}_{rqe} = \delta_{rq}. \qquad \ldots(18.9)$$

§ 19. The Irreducible Semi-normal Representations of $\mathcal{S}_n$

As a preliminary to the arguments of this section, we first explain what is meant by a matrix representation of a group or of a group algebra. If we can set up a correspondence

$$\tau_i : A_{\tau_i}$$

between the elements $\tau_1, \ldots, \tau_g$ of a group $\mathcal{G}$ and a set of square matrices $A_{\tau_1}, \ldots, A_{\tau_g}$, such that

$$A_{\tau_i} A_{\tau_j} = A_{\tau_i \tau_j},$$

then the matrices $A_{\tau_1}, \ldots, A_{\tau_g}$ are said to form a *matrix representation* of the group $\mathcal{G}$.

From the group $\mathcal{G}$ we can construct, just as we have done in § 6 for the group $\mathcal{S}_n$, a group algebra whose typical element is

$$X = \sum_i \xi_i \tau_i.$$

If we write

$$A_X \equiv \sum_i \xi_i A_{\tau_i},$$

then the correspondence

$$X : A_X$$

defines a matrix representation of the group algebra. The crucial feature of such a representation is that if X, Y be two elements of the group algebra, then

$$A_X A_Y = A_{XY}, \quad A_X \pm A_Y = A_{X \pm Y}.$$

A matrix representation $X : A_X$ is said to be *reducible* if a non-singular matrix H can be found, the same H for all X, such that for all X

$$H A_X H^{-1} = \begin{bmatrix} B_{11X} & O \\ B_{21X} & B_{22X} \end{bmatrix},$$

where the submatrices B_{11X} and B_{22X} are square, and O is a zero submatrix.

Since, with the notation of § 18,

$$\sum_{\alpha, r, s} u^{\alpha}_{rs(\tau_i \tau_j)} e^{\alpha}_{rs} = \tau_i \tau_j$$

$$= \sum_{\alpha, r, t} u^{\alpha}_{rt\tau_i} e^{\alpha}_{rt} \sum_{\beta, v, s} u^{\beta}_{vs\tau_j} e^{\beta}_{vs}$$

$$= \sum_{\alpha, r, t, s} u^{\alpha}_{rt\tau_i} u^{\alpha}_{ts\tau_j} e^{\alpha}_{rs},$$

and since the semi-normal units e^α_{rs} are linearly independent, we conclude that

$$u^\alpha_{rs(\tau_i\tau_j)} = \sum_t u^\alpha_{rt\tau_i} u^\alpha_{ts\tau_j}.$$

Thus, if $U^\alpha_{\tau_i}$ be the matrix of order $f^\alpha \times f^\alpha$ with elements $u^\alpha_{rs\tau_i}$, then

$$U^\alpha_{(\tau_i\tau_j)} = U^\alpha_{\tau_i} U^\alpha_{\tau_j}.$$

It follows from this that the matrices U^α_ϵ, $U^\alpha_{\tau_2}$, $U^\alpha_{\tau_3}$, . . . form a representation of $\mathcal{S}_n$, for these matrices have the same multiplication table as the permutations ϵ, τ_2, τ_3, . . . which they represent. In particular, it follows directly from equation (18.9) that $U^\alpha_\epsilon = I$, where I is the unit matrix of order f^α. Further, any substitutional expression $X \equiv \Sigma\, \lambda_i \tau_i$ can be represented by a matrix $U^\alpha_X \equiv \Sigma \lambda_i U^\alpha_{\tau_i}$.

Let us consider further the matrix U^α_X. Since

$$X = \sum_i \lambda_i \tau_i = \sum_i \lambda_i \sum_{\alpha, r, s} u^\alpha_{rs\tau_i} e^\alpha_{rs},$$

we find that the r,sth element u^α_{rsX} of the matrix U^α_X is given by

$$u^\alpha_{rsX} = \sum_i \lambda_i u^\alpha_{rs\tau_i}.$$

Since u^α_{rsX} is the coefficient of e^α_{rs} in X, it must also be the coefficient of e^α_{rs} in $e^\alpha_r X e^\alpha_s$. This fact will be used in § 21.

There is evidently one such representation associated with every partition α of n. It is called the *semi-normal representation* associated with α. If H be any non-singular matrix of order f^α, then the matrices

$$V^\alpha_{\tau_i} \equiv H U^\alpha_{\tau_i} H^{-1}$$

will also form a representation of $\mathcal{S}_n$, for

$$V^\alpha_{\tau_i} V^\alpha_{\tau_j} = H U^\alpha_{\tau_i} H^{-1} H U^\alpha_{\tau} H^{-1} = H U^\alpha_{(\tau_i\tau_j)} H^{-1} = V^\alpha_{(\tau_i\tau_j)}.$$

The representation $\tau_i : V^\alpha_{\tau_i}$ thus obtained is said to be *equivalent* to the representation $\tau_i : U^\alpha_{\tau_i}$. Certain of these equivalent representations will be discussed in the next chapter.

We have seen that any substitutional expression X can be expressed in the form

$$X = \sum_\alpha X^\alpha, \quad X^\alpha = \sum_{r,s} u^\alpha_{rsX} e^\alpha_{rs}.$$

Now, since

$$U^\alpha_X = H^{-1} V^\alpha_X H,$$

we have

$$u^\alpha_{rsX} = \sum_{p,q} h^{(-1)}_{rp} v^\alpha_{pqX} h_{qs},$$

where $h^{(-1)}_{rs}$, v^α_{rsX}, h_{rs} are the r,sth elements of the matrices H^{-1}, V^α_X, H respectively. It follows that

$$X^\alpha = \sum_{r,s,p,q} h^{(-1)}_{rp} v^\alpha_{pqX} h_{qs} e^\alpha_{rs}$$

$$= \sum_{p,q} v^\alpha_{pqX} \Big(\sum_{r,s} h^{(-1)}_{rp} e^\alpha_{rs} h_{qs}\Big).$$

Now if any particular matrix element v^α_{ijX} were zero for every expression X, then every X^α would be expressible as a linear combination of less than $(f^\alpha)^2$ expressions and consequently every X would be expressible as a linear combination of less than $\sum_\alpha (f^\alpha)^2$, that is $n!$, expressions. This is impossible, for the $n!$ permutations of $\mathcal{S}_n$ are linearly independent. It follows that it is not possible to find a non-singular matrix H such that for every X the matrix V^α_X takes the form

$$\begin{bmatrix} R_X & O \\ Q_X & W_X \end{bmatrix}.$$

In other words, the semi-normal representation and every other representation equivalent to it are *irreducible*.

The following theorem * is of some importance.

THEOREM 13. *If the matrix equation*

$$U^\alpha_X C = C U^\beta_X$$

holds for every substitutional expression X, *then*

(i) *if* $\alpha \neq \beta$, C *is a zero matrix ;*

(ii) *if* $\alpha = \beta$, C *is a scalar multiple of the unit matrix of order* f^α.

PROOF. If this equation holds for every X, then for every permutation τ

$$U^\alpha_\tau C = C U^\beta_\tau,$$

or

$$\sum_{s=1}^{f^\alpha} u^\alpha_{rs\tau} c_{st} = \sum_{q=1}^{f^\beta} c_{rq} u^\beta_{qt\tau}, \quad (r=1, \ldots, f^\alpha;\ t=1, \ldots, f^\beta).$$

Multiplying both sides by $u^\alpha_{vr\tau^{-1}}$ and summing over τ, we obtain

$$\sum_{s=1}^{f^\alpha} \sum_\tau u^\alpha_{rs\tau} u^\alpha_{vr\tau^{-1}} c_{st} = \sum_{q=1}^{f^\beta} \sum_\tau c_{rq} u^\beta_{qt\tau} u^\alpha_{vr\tau^{-1}}.$$

Using equation (18.7), we deduce that

$$\sum_s \delta_{sv} c_{st} \theta^\alpha = \sum_q c_{rq} \delta^{\alpha\beta} \delta_{qr} \delta_{vt} \theta^\alpha,$$

or

$$c_{vt} = \delta^{\alpha\beta} \delta_{vt} c_{rr}.$$

Hence if $\alpha \neq \beta$, C is the zero matrix, while if $\alpha = \beta$, C is a scalar multiple of the unit matrix.

This theorem implies amongst other things that no two different representations α and β are equivalent, that is, we cannot find a non-singular matrix C such that for every X

$$C^{-1} U^\alpha_X C = U^\beta_X.$$

* Schur, p. 410.

The content of this section may therefore be summarised in the following statement.

THEOREM 14. *Associated with every partition* α *of* n *there is an irreducible semi-normal matrix representation of* $\mathcal{S}_n$. *Two such representations associated with different partitions cannot be equivalent.*

§ 20. Expressions which do not Involve the Last Letter

We have seen that the f^α rows and f^α columns of a matrix U^α_X are associated with the f^α standard tableaux of shape α arranged according to the last letter sequence inasmuch as the row and column suffixes r, s refer to standard tableaux. Such a matrix can be partitioned naturally into submatrices according to the position of the last letter **n** in the tableaux which characterise the rows and columns. Thus in any submatrix arising from this partitioning all the rows are characterised by tableaux which have the letter **n** in the same position, and the same is true of the columns. In other words, all tableaux S characterising the rows of such a submatrix will have their S^* of the same shape. A similar result holds for the columns of the submatrix. A glance at the diagrams in § 28 will clarify this point.

Now if τ be any permutation which does not involve the last letter **n**, it can be expressed either as a linear combination of the semi-normal units e^β_{pq} of $\mathcal{S}_{n-1}$ or in terms of the units e^α_{rs} of $\mathcal{S}_n$. That is to say, we may write

$$\sum_{\alpha, r, s} u^\alpha_{rs\tau} e^\alpha_{rs} = \tau = \sum_{\beta, p, q} u^\beta_{pq\tau} e^\beta_{pq},$$

where in the first summation α ranges over the partitions of n and in the second summation β ranges over the partitions of $n-1$. Premultiply and postmultiply throughout by e^α_r and e^α_s respectively; then, using (17.4) and (17.1),

$$u^\alpha_{rs\tau} e^\alpha_{rs} = e^\alpha_r \tau e^\alpha_s = (1/\theta^\alpha)^2 \sum_{\beta, p, q} e^{\alpha *}_r E^\alpha_r e^{\alpha *}_r u^\beta_{pq\tau} e^\beta_{pq} e^{\alpha *}_s E^\alpha_s e^{\alpha *}_s.$$

But if $S^{\alpha *}_r$ and $S^{\alpha *}_s$ are of different shapes, that is if S^α_r and S^α_s have **n** in different positions, then by (17.4) $e^{\alpha *}_r e^\beta_{pq} e^{\alpha *}_s = 0$ and so $u^\alpha_{rs\tau} = 0$. It follows that all the submatrices of U^α_τ of the type under consideration which do not lie on the leading diagonal are zero. If, however, $S^{\alpha *}_r$ and $S^{\alpha *}_s$ are of the same shape, then S^α_r and S^α_s have **n** in the same position and so σ^α_{sr}, namely that permutation which changes S^α_r into S^α_s, does not involve **n**. This means that in this case σ^α_{sr} is also the permutation which changes $S^{\alpha *}_r$ into $S^{\alpha *}_s$. Thus $u^\alpha_{rs\tau}$, which by (18.4) is the coefficient of τ^{-1} in $\theta^\alpha e^\alpha_{sr}$ or in $e^{\alpha *}_s E^\alpha_s \sigma^\alpha_{sr} e^{\alpha *}_r$, is also the coefficient of τ^{-1} in $e^{\alpha *}_s E^{\alpha *}_s \sigma^\alpha_{sr} e^{\alpha *}_r$ or in $\theta^{\alpha *} e^{\alpha *}_{sr}$, and this, by (18.4), is precisely $u^{\alpha *}_{rs\tau}$; for, as was shown in § 17, those terms of E^α_s which are not included in $E^{\alpha *}_s$ all involve **n**. It follows that $u^\alpha_{rs\tau} = u^{\alpha *}_{rs\tau}$. This means that the submatrix of U^α_τ, which lies on the leading diagonal in the position corresponding to those tableaux for which S^* is of shape β, is precisely U^β_τ. In other words, if we use the notation of § 15,

$$U^\alpha_\tau = U^{[\alpha_k -]}_\tau \dotplus \dots \dotplus U^{[\alpha_1 -]}_\tau.$$

This result must also apply to any linear combination X of permutations which do not involve **n**. We therefore have the following result.†

THEOREM 15. *If X is any substitutional expression which does not involve the last letter* **n**, *then*

$$U_X^{\alpha} = U_X^{[\alpha_k -]} \dotplus \dots \dotplus U_X^{[\alpha_1 -]}.$$

For example, if X involves any or all of the letters **1, 2, . . ., 8** but not the letter **9**, then

$$U_X^{[4,2^2,1]} = \begin{bmatrix} U_X^{[4,2^2]} & \cdot & \cdot \\ \cdot & U_X^{[4,2,1^2]} & \cdot \\ \cdot & \cdot & U_X^{[3,2^2,1]} \end{bmatrix}.$$

It will be observed that the theorem just stated can be applied twice over if X involves neither **n** nor **n − 1**. Each $U_X^{[\alpha_i -]}$ is then a direct sum of submatrices $U_X^{[\alpha_i -, \alpha_j -]}$. This process can be carried to r stages if X does not involve any of the letters **n, n − 1, . . ., n − r + 1**.

§ 21. The Semi-normal Matrix for E_i^{α}

In the next section we shall require to know something about the matrix $U_{E_i^{\alpha}}^{\alpha}$, where S_i^{α} is a standard tableau. The element lying in the r,sth position of this matrix is (§ 19) the coefficient of e_{rs}^{α} in $e_r^{\alpha} E_i^{\alpha} e_s^{\alpha}$. Now, dropping the upper suffix α, we have

$$\begin{aligned} e_r E_i e_s &= (1/\theta)^2 e_r^* E_r e_r^* E_i e_s^* E_s e_s^* \\ &= (1/\theta)^2 (1/\theta_r^*)(1/\theta_s^*) e_r^* E_r e_r^{**} P_r^* N_r^* e_r^{**} P_i^* [\dots] N_i^* e_s^{**} P_s^* N_s^* e_s^{**} E_s e_s^*. \end{aligned}$$

The content of the vacant bracket is not relevant to our argument but its nature has been explained in § 17. Now by (14.2) the last expression vanishes if the shape of S_r^* comes before the shape of S_i^* or if the shape of S_i^* comes before the shape of S_s^*. This means that when U_{E_i} is partitioned according to the last letter sequence, all submatrices to the left of or below the submatrix containing the i,ith position are zero. In the same way the submatrix containing the i,ith position can itself be partitioned according to the position of the letter **n − 1**, and an argument similar to the preceding one will show that all the resulting submatrices to the left of or below the one containing the i,ith position are zero. Continuing this argument by applying it to the letters **n − 2, n − 3, . . .** we eventually find that every element on or below the leading diagonal of U_{E_i} is zero excepting possibly the element in the i,ith position. Since, however, $e_i E_i e_i = \theta e_i$, it is clear that this element has the value θ. We have now established the following result.

THEOREM 16. *The i,ith element of the matrix U_{E_i} is θ. All other elements lying on or below the leading diagonal are zero.*

† Young VI, p. 209.

§ 22. The Matrices $U^\alpha_{(\mathbf{n-1,n})}$

The permutation $(\mathbf{n-1}, \mathbf{n})$ commutes with every substitutional expression X formed from the letters $\mathbf{1}, \mathbf{2}, \ldots, \mathbf{n-2}$, and the same must be true of the matrices $U^\alpha_{(\mathbf{n-1,n})}$, U^α_X which represent these expressions. By using theorem 15 (§ 20) twice over it is clear that

$$U^\alpha_X = U^{k,k}_X \dotplus U^{k,k-1}_X \dotplus \ldots \dotplus U^{i,j}_X \dotplus \ldots \dotplus U^{j,i}_X \dotplus \ldots \dotplus U^{1,1}_X,$$

where i, j (previously written $[\alpha_i -, \alpha_j -]$) denotes the shape obtained from α by deleting the last spaces in rows i and j and where i, i is that obtained by deleting the last two spaces of row i, it being understood that any particular $U^{i,j}_X$ is absent if $[\alpha_i -]$ or if i, j is not a recognised shape. Evidently the shapes i, j and j, i are the same. It follows from this that the submatrices $U^{i,j}_X$ and $U^{j,i}_X$ are identical. On the other hand, we know from theorem 14 (§ 19) that the representations of $\mathcal{S}_{n-2}$ associated with different shapes are not equivalent, and that all such representations are irreducible.

Suppose now that $U^\alpha_{(\mathbf{n-1,n})}$ is partitioned like U^α_X according to the position of the last two letters. To specify the submatrices arising from this partitioning it will be advantageous to use double position indices. The position of the submatrix which appears in the i,jth row, that is, in the row associated with the partition $[\alpha_i -, \alpha_j -]$, and in the k,lth column will therefore be denoted by $(i, j;\ k, l)$.

Since the matrices $U^\alpha_{(\mathbf{n-1,n})}$ and U^α_X commute, we have

$$U^\alpha_{(\mathbf{n-1,n})} U^\alpha_X = U^\alpha_X U^\alpha_{(\mathbf{n-1,n})}.$$

Since U^α_X when partitioned takes a diagonal form, the last equation yields a set of equations, each of the type discussed in theorem 13 (§ 19), for the determination of the submatrices of $U^\alpha_{(\mathbf{n-1,n})}$. We therefore conclude from theorem 13 that the matrix $U^\alpha_{(\mathbf{n-1,n})}$ must have the following structure.

THEOREM 17. *The submatrices of $U^\alpha_{(\mathbf{n-1,n})}$ are zero matrices with the following exceptions :*

(i) *There is a submatrix cI, where c is numerical and I is a unit matrix, on the leading diagonal in each position $(i, i;\ i, i)$.* (The rows/columns of the submatrices in the i,ith row/column of the partitioned matrix are characterised by tableaux which have both $\mathbf{n-1}$ and $\mathbf{n}$ in the ith row.)

(ii) *There is also a submatrix cI on the leading diagonal in each position $(i, i-1;\ i, i-1)$ provided $\alpha_{i-1} = \alpha_i$.* This is the case in which i, j is, but $[\alpha_j -]$ is not, a recognised shape. (The rows/columns of the submatrices in the $i, i-1$th row/column of the partitioned matrix are characterised by tableaux which have both $\mathbf{n-1}$ and $\mathbf{n}$ in the α_ith column.)

(iii) *There are submatrices*

$c_{11}I$, $c_{12}I$	*in the positions*	$(i, j;\ i, j)$, $(i, j;\ j, i)$
$c_{21}I$, $c_{22}I$		$(j, i;\ i, j)$, $(j, i;\ j, i)$,

where c_{11}, c_{12}, c_{21}, c_{22} *are numerical and* I *is the unit matrix.* (The tableaux which characterise the rows/columns of the submatrices in the i,jth row/column of the partitioned matrix, have $n-1$ and n in different rows and different columns. They differ only from the tableaux associated with the j,ith row/column of the partitioned matrix in that the positions of $n-1$ and n are in each case interchanged.)

Our next object is to determine the numerical coefficients c, c_{11}, c_{12}, c_{21}, c_{22} with greater precision. Since $(n-1, n)^2 = \epsilon$, it follows that

$$(U^\alpha_{(n-1,\, n)})^2 = U^\alpha_\epsilon = I.$$

From this matrix equation we deduce the following relations :

$$c^2 = 1,$$

$$c_{11}{}^2 + c_{12}c_{21} = c_{22}{}^2 + c_{12}c_{21} = 1, \qquad \ldots (22.1)$$

$$c_{12}(c_{11} + c_{22}) = c_{21}(c_{11} + c_{22}) = 0. \qquad \ldots (22.2)$$

Further, $c_{21} = 1$, as we shall show by equating the matrices representing $\theta^\alpha e^\alpha_{rs}$ and $e^{\alpha*}_r E^\alpha_r \sigma^\alpha_{rs} e^{\alpha*}_s$, where S^α_r has n in the jth row and $n-1$ in the ith row $(i>j)$ and S^α_s is obtained from S^α_r by interchanging the letters $n-1$ and n. Thus $\sigma^\alpha_{rs} = (n-1, n)$ and s comes before r in the last letter sequence. Now it was shown in § 17 that $e^{\alpha*}_r = \Sigma\, e^\beta_t$ where the summation ranges over those tableaux S^β_t which yield the tableau $S^{\alpha*}_r$ on the deletion of the letter n. Of these tableaux only S^α_r is of shape α. It follows that the matrix $U^\alpha_{e^\alpha_r}$. has an element $+1$ in the rth place on the leading diagonal and that all the other elements are zero. A corresponding result holds for the matrix $U^\alpha_{e^\alpha_s}$. It follows now from theorems 16 (§ 21) and 17 that the matrix for $e^{\alpha*}_r E^\alpha_r$ (the product of the matrices for $e^{\alpha*}_r$ and E^α_r) and the matrix for $\sigma^\alpha_{rs} e^{\alpha*}_s$ (the product of the matrices for $(n-1, n)$ and $e^{\alpha*}_s$) are everywhere zero except in the positions indicated in the following respective diagrams :

$$\begin{bmatrix} & & \\ & \theta^\alpha\, ?\, \ldots\, ? & \\ & & \end{bmatrix} \quad \begin{matrix} \leftarrow \text{ row } s \rightarrow \\ \\ \leftarrow \text{ row } r \rightarrow \end{matrix} \quad \begin{bmatrix} & c_{11} & \\ & c_{21} & \\ & & \end{bmatrix}$$

$$\qquad\uparrow\qquad\uparrow \qquad\qquad\qquad\qquad\qquad \uparrow\qquad\uparrow$$
$$\text{col.}\quad\text{col.} \qquad\qquad\qquad\qquad \text{col.}\quad\text{col.}$$
$$\quad s\qquad r \qquad\qquad\qquad\qquad\qquad s\qquad r$$

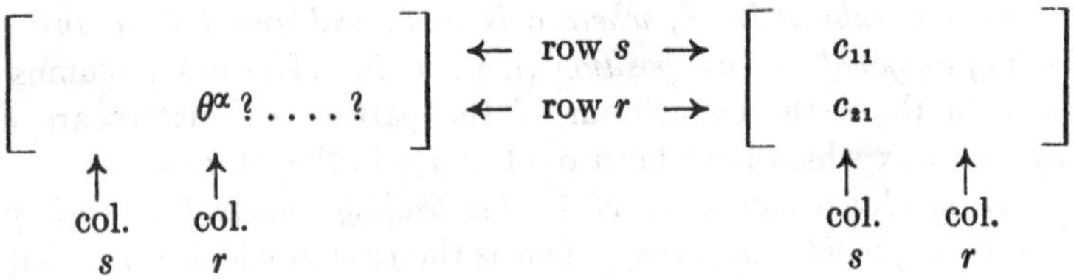

On forming the product of these two matrices, it is seen that the matrix for $e^{\alpha*}_r E^\alpha_r \sigma^\alpha_{rs} e^{\alpha*}_s$ has an element $\theta^\alpha c_{21}$ in the r,sth position and zero everywhere else. On the other hand, the matrix for $\theta^\alpha e^\alpha_{rs}$ has an element θ^α in the r,sth position and zero everywhere else. It follows that $c_{21} = +1$.

The equations (22.2) now yield

$$-c_{11} = +c_{22} = \rho,$$

say, while the equations (22.1) give

$$c_{12}=1-\rho^2.$$

We shall evaluate ρ in the next section but for the present we observe that the submatrices referred to in (iii) are †

$$\begin{matrix} -\rho I, (1-\rho^2)I & & (i,j\,;\,i,j), (i,j\,;\,j,i) \\ & \text{in the positions} & \\ I, \quad \rho I & & (j,i\,;\,i,j), (j,i\,;\,j,i). \end{matrix}$$

In the same way, if c be the r,rth element of $U^{\alpha}_{(\mathbf{n}-1,\mathbf{n})}$ and if S^{α}_{r} has $\mathbf{n-1}$ and $\mathbf{n}$ in the same row, then by considering the matrix equation representing the equality

$$e^{\alpha *}_{r}(\mathbf{n-1},\ \mathbf{n})E^{\alpha}_{r}e^{\alpha *}_{r}=e^{\alpha *}_{r}E^{\alpha}_{r}e^{\alpha *}_{r}=\theta^{\alpha}e^{\alpha}_{r}$$

we deduce with the aid of (17.7) and theorem 16 that $c=+1$. Thus the submatrix referred to in (i) is simply $+I$. Similarly, if S^{α}_{r} has $\mathbf{n-1}$ and $\mathbf{n}$ in the same column, the equality

$$e^{\alpha *}_{r}E^{\alpha}_{r}(\mathbf{n-1},\ \mathbf{n})e^{\alpha *}_{r}=-e^{\alpha *}_{r}E^{\alpha}_{r}e^{\alpha *}_{r}=-\theta^{\alpha}e^{\alpha}_{r}$$

leads to the conclusion that the submatrix referred to in (ii) is $-I$.

In passing we remark that no row and no column of the matrix $U^{\alpha}_{(\mathbf{n}-1,\mathbf{n})}$ contains more than two non-zero elements.

§ 23. The Matrices $U^{\alpha}_{(\mathbf{k}-1,\mathbf{k})}$

Before enunciating the culminating theorem of this chapter we must define the axial distance from a letter $\mathbf{p}$ to a letter $\mathbf{q}$ in a given tableau. Suppose that the letter $\mathbf{p}$ appears in the $i_{\mathbf{p}}$th row and in the $j_{\mathbf{p}}$th column of the tableau and that $\mathbf{q}$ appears in the $i_{\mathbf{q}}$th row and the $j_{\mathbf{q}}$th column. We define the *axial distance* from $\mathbf{p}$ to $\mathbf{q}$ to be

$$d_{\mathbf{p},\mathbf{q}}\equiv y_{\mathbf{q}}-y_{\mathbf{p}},$$

where $y_{\mathbf{p}}\equiv i_{\mathbf{p}}-j_{\mathbf{p}}$, $y_{\mathbf{q}}\equiv i_{\mathbf{q}}-j_{\mathbf{q}}$. According to our definition we have

$$d_{\mathbf{p},\mathbf{q}}=-d_{\mathbf{q},\mathbf{p}}.$$

The axial distance from $\mathbf{p}$ to $\mathbf{q}$ has a simple graphical interpretation. Starting from the position of $\mathbf{p}$ in the tableau we proceed by any rectangular route one space at a time until we reach the position of $\mathbf{q}$. Counting $+1$ for each step made downwards or to the left and -1 for each step made upwards or to the right, the resultant number of steps made will be the axial distance from $\mathbf{p}$ to $\mathbf{q}$.

THEOREM 18. *The matrix* $U^{\alpha}_{(\mathbf{k}-1,\mathbf{k})}$ *has*

(i) $+1$ *in the* r,rth *position if* S^{α}_{r} *has* $\mathbf{k-1}$ *and* $\mathbf{k}$ *in the same row ;*

(ii) -1 *in the* r,rth *position if* S^{α}_{r} *has* $\mathbf{k-1}$ *and* $\mathbf{k}$ *in the same column ;*

† Young VI, p. 215.

(iii) $-\rho,\ 1-\rho^2$ r,rth, r,sth

in the   *positions*

$1,\ \ +\rho$ s,rth, s,sth

if $r<s$ *and* S^α_s *is obtained from* S^α_r *by interchanging* $\mathbf{k-1}$ *and* $\mathbf{k}$, *where* $1/\rho$ *is the axial distance from* $\mathbf{k-1}$ *to* $\mathbf{k}$ *in* S^α_r ;

(iv) 0 *in all other positions.**

We observe that in every case the element in the r,rth position is the reciprocal of the axial distance from $\mathbf{k}$ to $\mathbf{k-1}$ in S^α_r.

The importance of theorem 18 lies in the fact that, as was proved in theorem 3 (§ 3), every permutation can be expressed as a product of transpositions of the form $(\mathbf{k-1, k})$. Since theorem 18 gives the matrices $U^\alpha_{(\mathbf{k-1,k})}$ in an explicit form, we have a method of writing down the semi-normal matrix U^α_τ representing any permutation τ. These in turn enable us to write down any matrix U^α_X, where X is any given substitutional expression.

Again, when the elements of all the matrices U^α_τ have been tabulated, it is a simple matter to read off from the table the expressions for the semi-normal units e^α_{rs} in terms of the permutations of $\mathcal{S}_n$. The matrix elements are of course the coefficients $u^\alpha_{rs\tau_i}$ which appear in equation (18.1). These are the same coefficients as appear in the equation (18.4), which give the semi-normal units in terms of the permutations.

PROOF. We shall prove this theorem by induction and it may be verified that it is true when $n=2,\ 3$. We can therefore assume its truth for matrices $U^\beta_{(\mathbf{k-1,k})}$, where β is a partition of $n-1$. It follows at once from theorem 15 (§ 20) that it is also true for partitions α of n provided $\mathbf{k}<\mathbf{n}$. It remains therefore to establish the truth of the theorem in the case $\mathbf{k}=\mathbf{n}$. This has been done in § 22 except for one fact, namely, we have still to show that in case (iii) $1/\rho$ is the axial distance from $\mathbf{n-1}$ to $\mathbf{n}$. We can do this by equating the s,rth elements of the matrix products

$$U_{(\mathbf{n-2,n-1})}U_{(\mathbf{n-1,n})}U_{(\mathbf{n-2,n-1})},\quad U_{(\mathbf{n-1,n})}U_{(\mathbf{n-2,n-1})}U_{(\mathbf{n-1,n})},\quad \ldots(23.1)$$

both of which are equal to $U_{(\mathbf{n-2,n})}$.

On account of the special structure of the matrices $U_{(\mathbf{n-2,n-1})}$ and $U_{(\mathbf{n-1,n})}$ it will be found that we need only consider the elements in the r,rth, r,sth, s,rth, s,sth positions of these matrices, for no other elements will contribute anything to the s,rth elements of the above products. To prove this for all possible cases would involve a rather long-winded explanation which we need not give here. In each case it is merely a matter of direct verification.

By the induction hypothesis the requisite elements of $U_{(\mathbf{n-2,n-1})}$ are

$$\begin{array}{r} \text{row } r \\ \text{row } s \\ \end{array} \begin{array}{c} \left[\begin{array}{cc} -\sigma & \cdot \\ \cdot & -\tau \end{array}\right], \\ \text{col. } r \quad \text{col. } s \end{array}$$

* Young VI, p. 217.

where $1/\sigma$ is the axial distance from $\mathbf{n-2}$ to $\mathbf{n-1}$ in S_r^α and $1/\tau$ is the axial distance from $\mathbf{n-2}$ to $\mathbf{n-1}$ in S_s^α or the axial distance from $\mathbf{n-2}$ to $\mathbf{n}$ in S_r^α. Further, we already know that the appropriate elements of $U^\alpha_{(\mathbf{n-1},\mathbf{n})}$ are

$$\begin{array}{c} \text{row } r \\ \text{row } s \end{array} \left[\begin{array}{cc} -\rho & 1-\rho^2 \\ 1 & \rho \end{array}\right]. \quad \begin{array}{cc} \text{col. } r & \text{col. } s \end{array}$$

Equating the s,rth elements of the matrix products (23.1) we obtain

$$\sigma\rho - \tau\rho = \tau\sigma.$$

Thus, if $d_{\mathbf{n-2},\mathbf{n-1}}$, $d_{\mathbf{n-2},\mathbf{n}}$, $d_{\mathbf{n-},\mathbf{1}\,\mathbf{n}}$ all refer to the tableau S_r^α,

$$\begin{aligned} \frac{1}{\rho} &= \frac{1}{\tau} - \frac{1}{\sigma} \\ &= d_{\mathbf{n-2},\mathbf{n}} - d_{\mathbf{n-2},\mathbf{n-1}} \\ &= (y_{\mathbf{n}} - y_{\mathbf{n-2}}) - (y_{\mathbf{n-1}} - y_{\mathbf{n-2}}) \\ &= (y_{\mathbf{n}} - y_{\mathbf{n-1}}) \\ &= d_{\mathbf{n-1},\mathbf{n}}, \end{aligned}$$

as required. This completes the proof of the theorem.

Two special cases of theorem 18 are of interest. Since $f^{[n]}$ and $f^{[1^n]}$ are both unity, the matrices $U_X^{[n]}$ and $U_X^{[1^n]}$ are of one row and one column. Now theorem 18 shows that for any $\mathbf{k}$ the matrix $U^{[n]}_{(\mathbf{k-1},\mathbf{k})} = +1$ and the matrix $U^{[1^n]}_{(\mathbf{k-1},\mathbf{k})} = -1$. It follows at once that for any permutation τ

$$U_\tau^{[n]} \equiv +1, \; U_\tau^{[1^n]} = \pm 1, \qquad \ldots(23.2)$$

according as τ is an even or an odd permutation.

We conclude this section with a deduction which will be used in § 37. Let τ be any permutation which can be expressed as a product of distinct transpositions of consecutive letters. Thus

$$\tau = \tau_1\tau_2 \ldots \tau_j,$$

where each τ_i is of the form $(\mathbf{k}_i - \mathbf{1},\ \mathbf{k}_i)$, and where $\mathbf{k}_1, \ldots, \mathbf{k}_j$ are all different. Then, according to the rules of matrix multiplication, the rth element on the leading diagonal of U_τ is just

$$\sum_{s,t,\ldots,p} u_{rs\tau_1} u_{st\tau_2} \ldots u_{pr\tau_j}.$$

We shall show that since $\tau_1, \ldots, \tau_j$ are distinct, the only non-zero term in the above summation is

$$u_{rr\tau_1} u_{rr\tau_2} \ldots u_{rr\tau_j}. \qquad \ldots(23.3)$$

Consider any term such as

$$u_{rs\tau_1} u_{st\tau_2} \ldots u_{pr\tau_j}. \qquad \ldots(23.4)$$

It follows that if $u_{rs\tau_i} \neq 0$, then either $r \neq s$ in which case, from theorem 18, $\sigma_{rs} = \tau_i$, or else $r = s$ in which case $\sigma_{rs} = \sigma_{rr} = \epsilon$. On the other hand, the product

$$\sigma_{rs}\sigma_{st} \ldots \sigma_{pr},$$

being equal to ϵ, cannot be expressed as a product of distinct τ_i. It follows that the only non-zero term (23.4) is that for which $r = s = t = \ldots = p$; that is to say (23.3) is the only non-zero term.

We have proved, therefore, that the rth diagonal element of U_τ is simply the product of the rth diagonal elements of the component factors U_{τ_i}. These, in turn, by theorem 18 are the quantities $1/d_{\mathbf{k}_i, \mathbf{k}_i - 1}$ evaluated for the tableau S_r. It is therefore a simple matter to write down the diagonal elements of any matrix U_τ, provided τ can be expressed as a product of distinct transpositions of the type $(\mathbf{k}-\mathbf{1}, \mathbf{k})$.

For example, if $\tau = (\mathbf{2}, \mathbf{3}, \mathbf{4})$ and $r = 2$,

$$S_2 = \begin{array}{|c|c|c|} \hline 1 & 2 & 4 \\ \hline 3 & 5 & \\ \cline{1-2} \end{array},$$

then $d_{4,3} = 3$ and $d_{3,2} = -2$. Since $(\mathbf{2}, \mathbf{3}, \mathbf{4}) = (\mathbf{2}, \mathbf{3})\,(\mathbf{3}, \mathbf{4})$, the 2nd diagonal element of U_τ must be $(1/d_{3,2})\,(1/d_{4,3})$, or $-\frac{1}{6}$.

CHAPTER IV

THE ORTHOGONAL AND NATURAL REPRESENTATIONS

Argument. Associated with every standard tableau S_r a certain numerical tableau function ψ_r is defined, and with the aid of this function the orthogonal representations of the symmetric group are derived from the semi-normal representations without difficulty. In this representation each matrix V_τ which represents a permutation τ, is an orthogonal matrix, and it follows from this that the associated invariant quadratic

$$\sum_\tau z'V'_\tau V_\tau z,$$

where z is a column vector whose components are the f variables * $z_1, \ldots, z_f$ and where the dash denotes transposition, takes the form

$$n!(z_1^2+\ldots+z_f^2).$$

Returning to the semi-normal representation, it is found that it is characterised by the invariant quadratic

$$n!\,(z_1^2/\psi_1+\ldots+z_f^2/\psi_f).$$

The second half of this chapter deals with the natural representation which was in fact the first to be discovered. This representation is intimately associated with a certain matrix Ξ^α, whose r,sth element is simply the coefficient of the unit permutation ϵ in E^α_{rs}.

§ 24. Equivalent Representations

In the last chapter it was shown that corresponding to any partition α of n it is possible to find a representation

$$\tau : U^\alpha_\tau$$

of the symmetric group $\mathcal{S}_n$. This representation is by no means unique nor was it the first to be discovered. If H be any non-singular matrix of order f^α, then, as we have seen in § 19, the matrices

$$V^\alpha_\tau \equiv H^{-1}U^\alpha_\tau H \qquad \ldots(24.1)$$

also give a representation

$$\tau : V^\alpha_\tau.$$

Although any two representations connected by the relation (24.1) are equiva-

* The $z_1, \ldots, z_f$ of this chapter should not be confused with $z_1, \ldots, z_n$ of § 1 and § 41.

lent,* the matrix elements and the units e^{α}_{rs} employed will be different in different equivalent representations. In this chapter we shall derive first the *orthogonal representation* and then the *natural representation*. Since we already have the semi-normal representation at hand it may be thought superfluous to describe the natural one, but in view of the important place it has had in the historical development of the subject we think it desirable to include some account of the natural representation also.

§ 25. The Invariant Quadratic

Let

$$\tau : A_\tau \qquad \ldots(25.1)$$

be any matrix representation of $\mathcal{S}_n$, where the matrices A_τ are non-singular and of order f. As is usual in matrix theory we shall use a dash to denote transposition. Thus A'_τ is the transposed matrix of A_τ. Let z be a variable column vector with components $z_1, \ldots, z_f$ and let Q be a square symmetric matrix of order f. Then the quadratic form

$$z'Qz$$

is called an *invariant quadratic* of the representation (25.1) if it is invariant under all transformations

$$z = A_\sigma y,$$

that is, if

$$A'_\sigma Q A_\sigma = Q \qquad \ldots(25.2)$$

for all σ. Now, by an elementary property of groups,

$$\sum_\tau A_\tau A_\sigma = \sum_\tau A_\tau,$$

from which it follows that

$$\sum_\tau z'A'_\tau A_\tau z = \sum_\tau y'A'_\sigma A'_\tau A_\tau A_\sigma y = \sum_\tau y'A'_\tau A_\tau y.$$

That is to say, $\sum_\tau z'A'_\tau A_\tau z$ is an invariant quadratic of the representation (25.1).

We shall now show that if a representation be an irreducible one, then its invariant quadratic is unique apart from a numerical factor. Suppose that the representation (25.1) is irreducible and that it has two invariant quadratics $z'Q_1z$ and $z'Q_2z$, where Q_2 is not a scalar multiple of Q_1. Then $z'(\lambda_1 Q_1 + \lambda_2 Q_2)z$ is also an invariant quadratic, where λ_1 and λ_2 are any numerical constants. It is therefore possible to choose λ_1 and λ_2 so that $\lambda_1 Q_1 + \lambda_2 Q_2$, which we may call Q, is singular and symmetric. We can therefore find † a non-singular matrix H such that

$$Q = H'RH,$$

where

$$R = \begin{bmatrix} I & \cdot \\ \cdot & \cdot \end{bmatrix}.$$

* The matrices are more often called "similar".

† T. A., p. 85.

Now the representation

$$\tau : B_\tau, \quad B_\tau \equiv HA_\tau H^{-1} \qquad \ldots(25.3)$$

is equivalent to (25.1). Also, since

$$A_\sigma = H^{-1}B_\sigma H, \quad A'_\sigma = H'B'_\sigma H'^{-1},$$

we can deduce from (25.2) that for all σ

$$H'B'_\sigma H'^{-1}H'RHH^{-1}B_\sigma H = H'RH.$$

It follows that for all σ

$$B'_\sigma RB_\sigma = R.$$

If we now partition B_σ in the same way as R, this last equation may be written

$$\begin{bmatrix} B'_{11\sigma} & B'_{21\sigma} \\ B'_{12\sigma} & B'_{22\sigma} \end{bmatrix} \begin{bmatrix} I & \cdot \\ \cdot & \cdot \end{bmatrix} \begin{bmatrix} B_{11\sigma} & B_{12\sigma} \\ B_{21\sigma} & B_{22\sigma} \end{bmatrix} = \begin{bmatrix} I & \cdot \\ \cdot & \cdot \end{bmatrix}.$$

Comparing the submatrices of the product on the left-hand side with the corresponding submatrices on the right, the following equations are obtained :

$$B'_{11\sigma}B_{11\sigma} = I, \quad B'_{11\sigma}B_{12\sigma} = 0, \quad B'_{12\sigma}B_{11\sigma} = 0, \quad B'_{12\sigma}B_{12\sigma} = 0.$$

From the first of these equations we see that $B'_{11\sigma}$ is non-singular. Since this is so, the second equation yields $B_{12\sigma} = 0$. Thus for all σ the matrix B_σ is of the form

$$B_\sigma = \begin{bmatrix} B_{11\sigma} & \cdot \\ B_{21\sigma} & B_{22\sigma} \end{bmatrix},$$

that is, the representation (25.3) is reducible (§ 19). It follows that the representation (25.1) is also reducible since it is equivalent to (25.3). This is contrary to our hypothesis, and consequently the invariant quadratic of an irreducible representation is unique apart from a numerical factor. We have, however, already found an invariant quadratic of the representation (25.1), so we may state that if the representation (25.1) is irreducible it has a unique invariant quadratic

$$\sum_\tau z'A'_\tau A_\tau z.$$

§ 26. The Tableau Function

Associated with any standard tableau S of n letters we first define a numerical function $\phi_S(\mathbf{n})$ such that if $\mathbf{n}$ appears in the first row of S, then $\phi_S(\mathbf{n}) \equiv 1$, but if $\mathbf{n}$ appears in any other row, then

$$\phi_S(\mathbf{n}) \equiv \prod_v (1 + c_v),$$

where there is one factor $1 + c_v$ for each row v of S which lies above the one in which $\mathbf{n}$ appears and where $1/c_v$ is the axial distance from the last letter of row v to the letter $\mathbf{n}$. We further define

$$\phi_S(\mathbf{n-1}) \equiv \phi_{S*}(\mathbf{n-1}), \quad \phi_S(\mathbf{n-2}) \equiv \phi_{S**}(\mathbf{n-2}), \ldots$$

and finally

$$\psi_S = \phi_S(\mathbf{n})\phi_S(\mathbf{n}-\mathbf{1}) \ldots \phi_S(\mathbf{1}).$$

It will be convenient also to write $\phi_r(\mathbf{t})$ and ψ_r in place of $\phi_{S_r}(\mathbf{t})$ and ψ_{S_r}. We call ψ_r the *tableau function* * of S_r.

Consider now any tableau S_r in which the letters $\mathbf{k}-\mathbf{1}$ and $\mathbf{k}$ appear in the ith and jth $(i<j)$ rows respectively and let the tableau S_s be defined by the relation $\sigma_{rs} = (\mathbf{k}-\mathbf{1}, \mathbf{k})$. That is, S_s has $\mathbf{k}-\mathbf{1}$ and $\mathbf{k}$ in the jth and ith rows respectively but is otherwise the same as S_r. A little consideration will show that unless $\mathbf{t} = \mathbf{k}-\mathbf{1}$ or $\mathbf{k}$

$$\phi_r(\mathbf{t}) = \phi_s(\mathbf{t}),$$

and that

$$\phi_r(\mathbf{k}-\mathbf{1}) = \phi_s(\mathbf{k}).$$

Further, $\phi_r(\mathbf{k})$ and $\phi_s(\mathbf{k}-\mathbf{1})$ differ in one factor only, namely the factor $1+c_i$. In the case of $\phi_r(\mathbf{k})$ this factor has the value $1+\dfrac{1}{\rho^{-1}}$, where $1/\rho$ is the axial distance from $\mathbf{k}-\mathbf{1}$ to $\mathbf{k}$ in S_r, but in $\phi_s(\mathbf{k}-\mathbf{1})$ it has the value $1+\dfrac{1}{\rho^{-1}-1}$, since the ith row of the tableau of $\mathbf{k}-\mathbf{1}$ letters which determines $\phi_s(\mathbf{k}-\mathbf{1})$ is one space shorter than the ith row of the tableau of k letters which determines $\phi_r(\mathbf{k})$. Thus

$$\phi_r(\mathbf{k}) \Big/ \left(1+\frac{1}{\rho^{-1}}\right) = \phi_s(\mathbf{k}-\mathbf{1}) \Big/ \left(1+\frac{1}{\rho^{-1}-1}\right),$$

or

$$\phi_r(\mathbf{k}) = (1-\rho^2)\phi_s(\mathbf{k}-\mathbf{1}).$$

It follows that if $\sigma_{rs} = (\mathbf{k}-\mathbf{1}, \mathbf{k})$ then

$$\psi_r = (1-\rho^2)\psi_s,$$

where $1/\rho$ is the axial distance from $\mathbf{k}-\mathbf{1}$ to $\mathbf{k}$ in S_r.

The following example should clarify the content of this section :

$$S_r = \begin{array}{|c|c|c|}\hline 1 & 2 & 3 \\ \hline 4 \\ \cline{1-1} 5 \\ \cline{1-1}\end{array}, \quad S_r^* = \begin{array}{|c|c|c|}\hline 1 & 2 & 3 \\ \hline 4 \\ \cline{1-1}\end{array}, \quad S_r^{**} = \begin{array}{|c|c|c|}\hline 1 & 2 & 3 \\ \hline\end{array},$$

$$S_r^{***} = \begin{array}{|c|c|}\hline 1 & 2 \\ \hline\end{array}, \quad S^{****} = \begin{array}{|c|}\hline 1 \\ \hline\end{array}\ ;$$

$$\phi_r(5) = (1+\tfrac{1}{1})(1+\tfrac{1}{4}), \quad \phi_r(4) = (1+\tfrac{1}{3}), \quad \phi_r(3) = 1, \quad \phi_r(2) = 1, \quad \phi_r(1) = 1\ ;$$

$$\psi_r = (1+\tfrac{1}{1})(1+\tfrac{1}{4})(1+\tfrac{1}{3}).$$

$$S_s = \begin{array}{|c|c|c|}\hline 1 & 2 & 4 \\ \hline 3 \\ \cline{1-1} 5 \\ \cline{1-1}\end{array}, \quad S_s^* = \begin{array}{|c|c|c|}\hline 1 & 2 & 4 \\ \hline 3 \\ \cline{1-1}\end{array}, \quad S_s^{**} = \begin{array}{|c|c|}\hline 1 & 2 \\ \hline 3 \\ \cline{1-1}\end{array},$$

$$S_s^{***} = \begin{array}{|c|c|}\hline 1 & 2 \\ \hline\end{array}, \quad S_s^{****} = \begin{array}{|c|}\hline 1 \\ \hline\end{array}\ ;$$

* This is not quite the same as Young's tableau function. Cf. Young VI, p. 221.

$$\phi_s(5)=(1+\tfrac{1}{1})(1+\tfrac{1}{4}),\quad \phi_s(4)=1,\quad \phi_s(3)=(1+\tfrac{1}{2}),\quad \phi_s(2)=1,\quad \phi_s(1)=1\,;$$

$$\psi_s=(1+\tfrac{1}{1})(1+\tfrac{1}{4})(1+\tfrac{1}{2}).$$

$$\frac{\psi_r}{\psi_s}=\frac{1+\frac{1}{3}}{1+\frac{1}{2}}=1-(\tfrac{1}{3})^2.\quad \sigma_{rs}=(3,4).$$

§ 27. The Orthogonal Representation

Restricting ourselves for the moment to a given partition α, let H and H^{-1} be the diagonal matrices

$$H=\begin{bmatrix}\sqrt{\psi_1} & & & \\ & \sqrt{\psi_2} & & \\ & & \ddots & \\ & & & \sqrt{\psi_f}\end{bmatrix},\qquad H^{-1}=\begin{bmatrix}\sqrt{\psi_1^{-1}} & & & \\ & \sqrt{\psi_2^{-1}} & & \\ & & \ddots & \\ & & & \sqrt{\psi_f^{-1}}\end{bmatrix}$$

and consider the representation

$$\tau: V_\tau,\quad V_\tau=H^{-1}U_\tau H, \qquad \ldots(27.1)$$

which is equivalent to the semi-normal representation $\tau: U_\tau$. A little reflection will show that $V_{(k-1,k)}$ will be the same as $U_{(k-1,k)}$ except that a submatrix

$$\begin{bmatrix}-\rho & 1-\rho^2\\ 1 & +\rho\end{bmatrix}$$

in $U_{(k-1,k)}$ will be replaced by

$$\begin{bmatrix}-\rho & (1-\rho^2)\sqrt{(\psi_s/\psi_r)}\\ \sqrt{(\psi_r/\psi_s)} & +\rho\end{bmatrix}$$

in $V_{(k-1,k)}$. But we have just shown in § 26 that $\psi_r/\psi_s=1-\rho^2$, so that the submatrix of $V_{(k-1,k)}$ in question is *

$$\begin{bmatrix}-\rho & \sqrt{(1-\rho^2)}\\ \sqrt{(1-\rho^2)} & +\rho\end{bmatrix}.$$

Since these submatrices are evidently orthogonal the same must be true of the matrix $V_{(k-1,k)}$. Thus, by theorem 3 (§ 3), any matrix V_τ is a product of orthogonal matrices and is therefore itself orthogonal. The representation (27.1) is called the *orthogonal representation* of $\mathfrak{S}_n$ associated with the partition α.

* Young VI, p. 218; Thrall, p. 612.

Since for all τ

$$V'_\tau V_\tau = V_\tau^{-1} V_\tau = I,$$

it is patent that the invariant quadratic

$$\sum_\tau z' V'_\tau V_\tau z$$

of this representation is simply

$$n!\, z'z,$$

that is *

$$n!\,(z_1^2 + \ldots + z_f^2).$$

Now, since $\sum_\tau V'_\tau V_\tau = n! I$ and since $H = H'$, it follows that

$$\sum_\tau H U'_\tau H^{-2} U_\tau H = n! I,$$

whence

$$\sum_\tau U'_\tau H^{-2} U_\tau = n! H^{-2}.$$

But

$$\sum_\tau z' U'_\tau H^{-2} U_\tau z$$

is an invariant quadratic of the representation $\tau : U_\tau$ since it is unaltered by any transformation $z = U_\sigma y$; also, since the representation $\tau : U_\tau$ is irreducible, this quadratic is unique (§ 25). It follows that the invariant quadratic of the semi-normal representation is

$$n! z' H^{-2} z,$$

or †

$$n!\,(z_1^2/\psi_1 + \ldots + z_n^2/\psi_n).$$

Since $V_\tau = H^{-1} U_\tau H$, it follows that

$$v_{rs\tau} = \sqrt{(\psi_s/\psi_r)}\, u_{rs\tau},$$

where $v_{rs\tau}$ is the r,sth element of the matrix V_τ. So

$$\tau = \sum_{\alpha, r, s} u^\alpha_{rs\tau}\, e^\alpha_{rs} = \sum_{\alpha, r, s} v^\alpha_{rs\tau} o^\alpha_{rs},$$

where

$$o^\alpha_{rs} \equiv \sqrt{(\psi^\alpha_r/\psi^\alpha_s)}\, e^\alpha_{rs}.$$

The expressions o^α_{rs} are called the *orthogonal units* and are in general irrational expressions. It is easily verified that they satisfy the equations

$$o^\alpha_{rs} o^\beta_{pq} = \delta^{\alpha\beta} \delta_{sp} o^\alpha_{rq},$$

$$\sum_{\alpha, r} o^\alpha_{rr} = \epsilon,$$

$$o^\alpha_{rs} = (1/\theta^\alpha) \sum_\tau v^\alpha_{sr\tau^{-1}} \tau,$$

* Young VI, p. 218. † Young VI, p. 217.

which are analogous to (17.4), (17.3) and (18.4). In particular, it should be observed that

$$o_{rr}^{\alpha}=e_{rr}^{\alpha}. \qquad \ldots(27.2)$$

§ 28. The Matrix Ξ^{α}

In the case $n \leqslant 4$ it is found that the formulae (i), (ii), (iii) of theorem 11 (§ 14), applied to standard tableaux only, are comprised in the single formula

$$E_{rs}^{\alpha}E_{uv}^{\beta}=\delta^{\alpha\beta}\delta_{su}\theta^{\alpha}E_{rv}^{\alpha}.$$

This is merely a matter of direct verification. In such a case, therefore, the $n!$ expressions

$$g_{rs}^{\alpha}\equiv(1/\theta^{\alpha})E_{rs}^{\alpha}, \quad (n \leqslant 4)$$

could be chosen as units in place of the semi-normal units e_{rs}^{α} or the orthogonal units o_{rs}^{α}. They are called the *natural units* of the group $\mathfrak{S}_n$. The comparative simplicity of their definition might lead one to expect a corresponding result for $n>4$. Unfortunately this is not so. Unfortunately, because it is a comparatively easy matter by means of theorem 7 (§ 11) to evaluate the coefficient of any τ in E_{rs}^{α} but not nearly so easy to find the coefficient of any τ in the units g_{rs}^{α} which we shall presently define for all values of n.

We begin by defining numbers ξ_{us}^{α} such that

$$E_{rs}^{\alpha}E_{uv}^{\beta}=\delta^{\alpha\beta}\theta^{\alpha}\xi_{us}^{\alpha}E_{rv}^{\alpha}. \qquad \ldots(28.1)$$

Comparing this with the formulae (i), (ii), (iii) of theorem 11 (§ 14), we deduce that

(i) $\xi_{us}=0$ if $N_sP_u=0$, or, by theorem 7 (§ 11), if the coefficient of ϵ in E_{us} is 0.

(ii) $\xi_{us}=+1$ if $N_sP_u(=N_tP_t)\neq 0$ and σ_{st} is even, in which case the coefficient of ϵ in E_{us} is $+1$.

(iii) $\xi_{us}=-1$ if $N_sP_u(=N_tP_t)\neq 0$ and σ_{st} is odd, in which case the coefficient of ϵ in E_{us} is -1.

In all cases, therefore, ξ_{us} is simply the coefficient of ϵ in E_{us}.

The matrix Ξ^{α} with ξ_{rs}^{α} in the r,sth position can now be constructed according to the method of theorem 7 (§ 11). For example, if $\alpha=[4, 2]$ there are nine standard tableaux which are, when arranged according to the last letter sequence,

1234	**1235**	**1245**	**1345**	**1236**	**1246**	**1346**	**1256**	**1356**
56	**46**	**36**	**26**	**45**	**35**	**25**	**34**	**24**

From the following table, which consists of 9^2 diagrams of the type described in § 10.

	1234 56	1235 46	1245 36	1345 26	1236 45	1246 35	1346 25	1256 34	1356 24
1234 56	1234 56	.	.	.	.	.	.	.	.
1235 46	.	1235 46	.	.	.	.	.	.	.
1245 36	.	.	1245 36	.	.	.	.	.	.
1345 26	.	.	.	1345 26	.	.	.	.	.
1236 45	.	.	.	.	1236 45	.	.	.	.
1246 35	.	.	.	.	.	1246 35	.	.	.
1346 25	1634 52	.	.	.	.	.	1346 25	.	.
1256 34	.	.	.	.	.	.	.	1256 34	.
1356 24	.	1635 42	.	.	1536 42	.	.	.	1356 24

we see that

$$\Xi^{[4,2]} = \left[\begin{array}{cccc|ccccc} 1 & . & . & . & . & . & . & . & . \\ . & 1 & . & . & . & . & . & . & . \\ . & . & 1 & . & . & . & . & . & . \\ . & . & . & 1 & . & . & . & . & . \\ \hline . & . & . & . & 1 & . & . & . & . \\ . & . & . & . & . & 1 & . & . & . \\ -1 & . & . & . & . & . & 1 & . & . \\ . & . & . & . & . & . & . & 1 & . \\ . & -1 & . & . & -1 & . & . & . & 1 \end{array}\right]$$

Thus $\xi_{7,1} = -1$ because the permutation (26) which changes $\begin{matrix}1234\\56\end{matrix}$ into $\begin{matrix}1634\\52\end{matrix}$ is odd.

It will be observed that in the example cited all the elements above the leading diagonal vanish. This is not an accident but is true in general, as we shall now prove by induction. To do this we partition the rows and columns of Ξ according to the position of the last letter, as is indicated in the example by means of dotted lines. Any submatrix lying on the leading diagonal is precisely a matrix Ξ^{β} for some partition β of $n-1$ and therefore by the induction hypothesis has the required property. In the example the two such submatrices are $\Xi^{[4,1]}$ and $\Xi^{[3,2]}$. Consider now any submatrix lying above the leading diagonal and let ξ_{rs} $(r<s)$ be any element of it. Then S_r and S_s have **n** in different rows so that S_r^* and S_s^*, the tableaux obtained from S_r and S_s by deleting the last letter **n**, are of different shapes, say β and γ respectively. Further $\beta>\gamma$ since $r<s$. Hence, by (14.1),

$$N_s^* P_r^* = P_r^* N_s^* = 0.$$

This implies that some row of S_r^* contains two letters which both lie in the same column of S_s^*. It follows that some row of S_r contains two letters which both lie in the same column of S_s. It should now be clear from the schematic method of calculating the matrix Ξ that $\xi_{rs}=0$. The cases $n=2$, 3, 4 form a basis of the induction which is now proved. Adding to the above the fact that every ξ_{rr} on the leading diagonal of Ξ must be $+1$, we have the following result.

THEOREM 19. *Ξ is a triangular matrix with determinant $+1$. Every element on the leading diagonal is $+1$ and every element above this diagonal is zero.*

§ 29. The Natural Representation

It is an immediate consequence of theorem 19 that the matrix Ξ is non-singular and that its inverse Ξ^{-1} exists. If we denote the r,sth element of Ξ^{-1} by η_{rs}, then

$$\sum_t \xi_{rt}\eta_{ts} = \delta_{rs}. \qquad \ldots(29.1)$$

Let g_{rs}^{α} be defined by the matrix equation *

$$g^{\alpha} = (1/\theta^{\alpha})E^{\alpha}(\Xi^{\alpha})^{-1}, \qquad \ldots(29.2)$$

that is,

$$g_{rs}^{\alpha} = (1/\theta^{\alpha})\sum_t E_{rt}^{\alpha}\eta_{ts}^{\alpha}. \qquad \ldots(29.3)$$

Then by (28.1) and (29.1)

$$\begin{aligned} g_{rs}^{\alpha}g_{uv}^{\beta} &= (1/\theta^{\alpha}\theta^{\beta})\sum_{t,w} E_{rt}^{\alpha}\eta_{ts}^{\alpha}E_{uw}^{\beta}\eta_{wv}^{\beta} \\ &= (1/\theta^{\alpha})\delta^{\alpha\beta}\sum_{t,w} E_{rw}^{\alpha}\xi_{ut}^{\alpha}\eta_{ts}^{\alpha}\eta_{wv}^{\alpha} \\ &= (1/\theta^{\alpha})\delta^{\alpha\beta}\delta_{us}\sum_{w} E_{rw}^{\alpha}\eta_{wv}^{\alpha} \\ &= \delta^{\alpha\beta}\delta_{us}g_{rv}^{\alpha} \end{aligned} \qquad \ldots(29.4)$$

in analogy with (17.4).†

* Young IV, p. 259.

† Young III, p. 264.

Since

$$E_{rt}^{\alpha}=E_{r}^{\alpha}\sigma_{rt}^{\alpha}=E_{r}^{\alpha}\sigma_{rs}^{\alpha}\sigma_{st}^{\alpha}=E_{rs}^{\alpha}\sigma_{st}^{\alpha},$$

the equation (29.3) may be written *

$$g_{rs}^{\alpha}=(1/\theta^{\alpha})E_{rs}^{\alpha}M_{s}^{\alpha}, \qquad \ldots(29.5)$$

where

$$M_{s}^{\alpha}\equiv\sum_{t}\eta_{ts}^{\alpha}\sigma_{st}^{\alpha}.$$

The equation (29.5) is equivalent to Young's original definition of the natural units g_{rs}^{α}.

By comparing the coefficients of ϵ in the corresponding elements of the matrices on both sides of equation (29.2) and recalling that ξ_{ru}^{α} is merely the coefficient of ϵ in E_{ru}^{α}, it is patent that the coefficient of ϵ in g_{rs}^{α} is the r,sth element of the matrix

$$(1/\theta^{\alpha})\Xi^{\alpha}(\Xi^{\alpha})^{-1},$$

which is simply the matrix $(1/\theta^{\alpha})I$. In other words, the coefficient of ϵ in g_{rs}^{α} is $(1/\theta^{\alpha})\delta_{rs}$. From this we see that no g_{rs}^{α} vanishes, for

$$g_{rs}^{\alpha}g_{sr}^{\alpha}=g_{rr}^{\alpha}$$

and the coefficient of ϵ on the right-hand side does not vanish.

Since Ξ^{α} is of the nature stated in theorem 19 (§ 28) and since

$$g^{\alpha}\Xi^{\alpha}=(1/\theta^{\alpha})E^{\alpha},$$

it follows that

$$g_{ff}^{\alpha}=(1/\theta^{\alpha})E_{ff}^{\alpha}. \qquad \ldots(29.6)$$

In view of these results, the $n!$ natural units g_{rs}^{α} like the semi-normal units are linearly independent and any substitutional expression is a linear combination of them.† The arguments of § 18 hold good when applied to the natural units and, in particular, the coefficient of g_{rs}^{α} in any permutation τ is θ^{α} times the coefficient of τ^{-1} in g_{sr}^{α}. Since the coefficient of ϵ in g_{sr}^{α} is $(1/\theta^{\alpha})\delta_{rs}$, it follows that the coefficient of g_{rs}^{α} in ϵ is δ_{rs}. Thus

$$\epsilon=\sum_{\alpha,r}g_{rr}^{\alpha},$$

in analogy with (17.3).

As might be expected, we are also able to show that

$$\sum_{r}g_{rr}^{\alpha}=\sum_{r}e_{rr}^{\alpha} \qquad \ldots(29.7)$$

although in general $g_{rr}^{\alpha}\neq e_{rr}^{\alpha}$. To do this, we first observe that E_{rt}^{α} can be expressed in the form

$$E_{rt}^{\alpha}=\sum_{u,v,\beta}\lambda_{uv}^{\beta}e_{uv}^{\beta},$$

* Young III, p. 263.

† Young III, p. 255.

where the coefficients λ^{β}_{uv} are numerical. It follows that

$$e^{\beta}_{u}E^{\alpha}_{rt}e^{\beta}_{v}=\lambda^{\beta}_{uv}e^{\beta}_{uv}.$$

Now

$$e^{\beta}_{u}E^{\alpha}_{rt}e^{\beta}_{v}=(1/\theta^{\beta})^{2}e^{\beta *}_{u}E^{\beta}_{uu}e^{\beta *}_{u}E^{\alpha}_{rt}e^{\beta *}_{v}E^{\beta}_{vv}e^{\beta *}_{v}$$

and by theorem 10b (§ 14) the right-hand side of this equation vanishes if $\alpha \neq \beta$. We conclude therefore that

$$E^{\alpha}_{rt}=\sum_{u,v}\lambda^{\alpha}_{uv}e^{\alpha}_{uv}.$$

Combining this last result with equation (29.3), it is clear that $\sum_{r} g^{\alpha}_{rr}$ can be written in the form

$$\sum_{r} g^{\alpha}_{rr}=\sum_{uv}\mu^{\alpha}_{uv}e^{\alpha}_{uv},$$

where the coefficients μ^{α}_{uv} are numerical. Hence

$$\sum_{s} e^{\alpha}_{ss}=\epsilon\sum_{s} e^{\alpha}_{ss}=\sum_{\beta,r} g^{\beta}_{rr}\sum_{s} e^{\alpha}_{ss}=\sum_{\beta,u,v}\mu^{\beta}_{uv}e^{\beta}_{uv}\sum_{s} e^{\alpha}_{ss}=\sum_{u,s}\mu^{\alpha}_{us}e^{\alpha}_{us}=\sum_{r} g^{\alpha}_{rr},$$

which is what we set out to prove.

These *natural units* g^{α}_{rs} lead to the *natural representation* of $\mathcal{S}_n$ just as the semi-normal units e^{α}_{rs} led to the semi-normal representation. The matrix elements of the natural representation are the coefficients of the various permutations in the expression $\sum_{t} E^{\alpha}_{rt}\eta^{\alpha}_{ts}$ which appears on the right-hand side of (29.3). These elements are therefore integers. We shall not make much further use of the natural representation as for most purposes the semi-normal representation is equally suitable and often more apposite. In § 31 we shall establish the equivalence of the natural and semi-normal representations.

§ 30. Expressions for the Units g^{α}_{ii}

Certain results * due to H. W. Turnbull shed an interesting light on the nature of the natural units g^{α}_{rs}. Confining our attention for the present to one shape α, we may drop this suffix whenever it appears. Let us write

$$Q_i=\sum_{k} g_{kk}-(1/\theta)E_{ii}.$$

Since the matrix equation (29.2) may be written in the form

$$(1/\theta)E=g\,\Xi,$$

and since the matrix Ξ has the form described in theorem 19 (§ 28), we conclude that

$$(1/\theta)E_{ii}=g_{ii}+g_{i,i+1}\xi_{i+1,i}+\ldots+g_{if}\,\xi_{fi}$$

and hence that the natural matrix for $(1/\theta)E_{ii}$ is of the form

* Suggested orally to the author.

$$\begin{bmatrix} 0 & & & & & & \\ & \ddots & & & & & \\ & & 0 & & & & \\ & & & 1 & \xi_{i+1,\,i} & \cdots & \xi_{fi} \\ & & & & 0 & & \\ & & & & & \ddots & \\ & & & & & & 0 \end{bmatrix} \leftarrow \text{row } i,$$

all elements not otherwise indicated being zero. Consequently, since $\sum_k g_{kk}$ is represented by the unit matrix, the natural matrix for Q_i is of the form

$$\begin{bmatrix} 1 & & & & & & \\ & \ddots & & & & & \\ & & 1 & & & & \\ & & & 0 & -\xi_{i+1,\,i} & \cdots & -\xi_{fi} \\ & & & & 1 & & \\ & & & & & \ddots & \\ & & & & & & 1 \end{bmatrix} \leftarrow \text{row } i.$$

It will now be shown by induction that the product

$$Q_i Q_{i+1} \ldots Q_f$$

is represented by the diagonal matrix

$$\begin{bmatrix} 1 & & & & & \\ & \ddots & & & & \\ & & 1 & & & \\ & & & 0 & & \\ & & & & 0 & \\ & & & & & \ddots \\ & & & & & & 0 \end{bmatrix} \leftarrow \text{row } i.$$

If we assume that this formula holds for the product $Q_{i+1} \ldots Q_f$, the ordinary rules of matrix multiplication show that the product of the matrices for Q_i

and for $Q_{i+1} \ldots Q_f$ is again of the required form. The basis of this induction is the fact that the matrix for Q_f is

$$\begin{bmatrix} 1 & & & \\ & \ddots & & \\ & & 1 & \\ & & & 0 \end{bmatrix}.$$

This follows at once from the result $g_{ff} = (1/\theta)E_{ff}$ which has already been obtained. The induction is therefore established.

The result just proved demonstrates that the product $Q_1Q_2 \ldots Q_f$ is represented by the zero matrix. It also shows that the product of the matrices representing $(1/\theta)E_{ii}$ and $Q_{i+1} \ldots Q_f$ is the matrix with $+1$ in the i,ith position and zero elsewhere, that is, the matrix representing g_{ii}. Since all the expressions under consideration are linear combinations of the g_{rs} associated with one partition α only, we conclude that

$$g_{ii} = (1/\theta)E_{ii}Q_{i+1} \ldots Q_f$$

and that

$$Q_1Q_2 \ldots Q_f = 0.$$

Consider now the expressions

$$(1/\theta^\alpha)E_{ii}^\alpha R_{i+1}^\alpha \ldots R_f^\alpha$$

and

$$H^\alpha \equiv R_1^\alpha R_2^\alpha \ldots R_f^\alpha,$$

where

$$R_j^\alpha \equiv \epsilon - (1/\theta^\alpha)E_{jj}^\alpha.$$

We have already proved that $\epsilon = \sum_{\beta, t} g_{tt}^\beta$. Since the factor E_{ii}^α is a linear combination of the units g_{rs}^α, it annihilates every g_{rs}^β for which $\beta \neq \alpha$. It follows that

$$(1/\theta^\alpha)E_{ii}^\alpha R_{i+1}^\alpha \ldots R_f^\alpha = (1/\theta^\alpha)E_{ii}^\alpha Q_{i+1}^\alpha \ldots Q_f^\alpha = g_{ii}^\alpha.$$

Written out in full this formula reads

$$g_{ii}^\alpha = (1/\theta^\alpha)E_{ii}^\alpha \prod_{j=i+1}^{f} (\epsilon - (1/\theta^\alpha)E_{jj}^\alpha). \qquad \ldots(30.1)$$

Similarly, the only terms in H^α which are not annihilated in view of (29.4) are

$$\sum_{\beta \neq \alpha} \sum_t (g_{tt}^\beta)^f - Q_1^\alpha \ldots Q_f^\alpha.$$

Since we have just proved that the second term is zero, these reduce to $\sum_{\beta \neq \alpha} \sum_t g_{tt}^\beta$, or

$$\epsilon - \sum_t g_{tt}^\alpha.$$

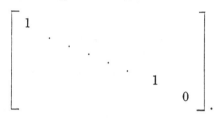

Thus H^α contains no units g^α_{rs}, H^β no units g^β_{rs} and so on. The product $\prod_\alpha H^\alpha$ must therefore vanish since it cannot involve any of these units. In other words,

$$\prod_\alpha \prod_{j=1}^{f} (\epsilon - (1/\theta^\alpha) E^\alpha_{jj}) = 0. \qquad \ldots(30.2)$$

In illustration of these results we refer to the case $n=3$, in which there are four standard tableaux,

$$S^{[3]}_1 = \begin{array}{|c|c|c|}\hline 1 & 2 & 3 \\ \hline\end{array}, \quad S^{[2,1]}_1 = \begin{array}{|c|c|}\hline 1 & 2 \\ \hline 3 \\ \cline{1-1}\end{array}, \quad S^{[2,1]}_2 = \begin{array}{|c|c|}\hline 1 & 3 \\ \hline 2 \\ \cline{1-1}\end{array}, \quad S^{[1^3]}_1 = \begin{array}{|c|}\hline 1 \\ \hline 2 \\ \hline 3 \\ \hline\end{array}.$$

The formulae (30.1) and (30.2) show that

$$g^{[3]}_{11} = \tfrac{1}{6} E^{[3]}_{11}, \quad g^{[2,1]}_{11} = \tfrac{1}{3} E^{[2,1]}_{11} (\epsilon - \tfrac{1}{3} E^{[2,1]}_{22}),$$

$$g^{[2,1]}_{22} = \tfrac{1}{3} E^{[2,1]}_{22}, \quad g^{[1^3]}_{11} = \tfrac{1}{6} E^{[1^3]}_{11},$$

$$(\epsilon - \tfrac{1}{6} E^{[3]}_{11})(\epsilon - \tfrac{1}{3} E^{[2,1]}_{11})(\epsilon - \tfrac{1}{3} E^{[2,1]}_{22})(\epsilon - \tfrac{1}{6} E^{[1^3]}_{11}) = 0.$$

The other units g^α_{rs} are easily obtained once the units g^α_{ss} are known, for by (29.5)

$$g^\alpha_{rs} = (1/\theta^\alpha) E^\alpha_{rs} M^\alpha_s = (1/\theta^\alpha) \sigma^\alpha_{rs} E^\alpha_{ss} M^\alpha_s = \sigma^\alpha_{rs} g^\alpha_{ss}.$$

It may be worth emphasising here that

$$g^\alpha_{rs} \neq g^\alpha_{rr} \sigma^\alpha_{rs}.$$

§ 31. Equivalence of the Natural and Semi-normal Representations

The argument used in § 29 can be extended without difficulty to show that not only g^α_{rr} but also every g^α_{ij} is a linear function of the semi-normal units e^α_{kl} only, α remaining fixed. Thus, dropping the suffix α, we may write

$$g_{ij} = \sum_{k,l} c_{ij,kl} e_{kl}, \qquad \ldots(31.1)$$

where the coefficients $c_{ij,kl}$ are numerical. Now, since

$$g_{ij} g_{rs} = \delta_{jr} g_{is}, \quad e_{ij} e_{rs} = \delta_{jr} e_{is},$$

it follows that

$$\begin{aligned} \delta_{jr} \sum_{k,v} c_{is,kv} e_{kv} &= \delta_{jr} g_{is} \\ &= g_{ij} g_{rs} \\ &= \sum_{k,l} c_{ij,kl} e_{kl} \sum_{t,v} c_{rs,tv} e_{tv} \\ &= \sum_{k,l,t,v} c_{ij,kl} c_{rs,tv} \delta_{lt} e_{kv} \\ &= \sum_{k,t,v} c_{ij,kt} c_{rs,tv} e_{kv}. \end{aligned}$$

Whence, comparing coefficients,

$$\delta_{jr} c_{is,\,kv} = \sum_t c_{ij,\,kt} c_{rs,\,tv}. \qquad \ldots(31.2)$$

Again, any linear function $\sum_{ij} w_{ij} g_{ij}$ of the units g_{ij} is also a linear function $\sum_{k,t} u_{kt} e_{kt}$ of the units e_{kt}. From (31.1) we readily deduce that

$$\sum_{i,j,k,t} w_{ij} c_{ij,\,kt} e_{kt} = \sum_{k,t} u_{kt} e_{kt},$$

and so

$$\sum_{i,j} w_{ij} c_{ij,\,kt} = u_{kt}. \qquad \ldots(31.3)$$

If we now multiply both sides of this equation by $c_{rs,\,tv}$ and sum over t, we get

$$\sum_{i,j,t} w_{ij} c_{ij,\,kt} c_{rs,\,tv} = \sum_t u_{kt} c_{rs,\,tv}.$$

Applying (31.2) to this last equation, we find that

$$\sum_i w_{ir} c_{is,\,kv} = \sum_t u_{kt} c_{rs,\,tv}.$$

Similarly, by multiplying both sides of (31.3) by $c_{rs,\,vk}$ and summing over k we get

$$\sum_j w_{sj} c_{rj,\,vt} = \sum_k u_{kt} c_{rs,\,vk}.$$

These last two equations may be written in matrix form. If W_X, U_X, $C_{.s.v}$, $C_{r.v.}$ denote the matrices, each of order f, whose i,jth elements are respectively w_{ij}, u_{ij}, $c_{is,\,jv}$, $c_{ri,\,vj}$, then the last two equations take the form

$$W'_X C_{.s.v} = C_{.s.v} U'_X, \qquad \ldots(31.4)$$
$$W_X C_{r.v.} = C_{r.v.} U_X,$$

where the dash as usual denotes transposition.

Now the matrices W_X and U_X may be regarded as the natural and semi-normal matrices associated with a given partition, which represent some substitutional expression X. Further, the matrices $C_{.s.v}$ and $C_{r.v.}$ are independent of the choice of X. Since these matrices are not zero for every s and v or for every r and v, it follows from Schur's theorem, which is proved below, that the natural and semi-normal representations associated with the same partition are equivalent.

THEOREM 20 (Schur's Theorem *). *If A_X and B_X are the matrices for a substitutional expression X corresponding to two irreducible representations, and if for every X*

$$QA_X = B_X Q,$$

then either the representations are equivalent, or Q is the zero matrix.

We require this theorem only for the case where the two representations

* Schur, p. 409.

are of the same order f and we shall confine our proof to this case. The reader will be aware of the close affinity of this theorem to theorem 13 (§ 19).

PROOF. Suppose that the matrix Q is of rank r $(0<r<f)$. It is known that non-singular matrices H and K can be found such that *

$$HQK=\begin{bmatrix} I & \cdot \\ \cdot & \cdot \end{bmatrix},$$

where I is the unit matrix of order r. It follows that

$$\begin{bmatrix} I & \cdot \\ \cdot & \cdot \end{bmatrix} K^{-1}A_XK=HB_XH^{-1}\begin{bmatrix} I & \cdot \\ \cdot & \cdot \end{bmatrix}. \qquad \ldots(31.5)$$

The matrices $K^{-1}A_XK$ and HBH^{-1} can now be partitioned in the same way as HQK and we can write

$$K^{-1}A_XK\equiv\begin{bmatrix} L_{11} & L_{12} \\ L_{21} & L_{22} \end{bmatrix}, \quad HB_XH^{-1}\equiv\begin{bmatrix} M_{11} & M_{12} \\ M_{21} & M_{22} \end{bmatrix}.$$

The equation (31.5) now yields

$$\begin{bmatrix} L_{11} & L_{12} \\ \cdot & \cdot \end{bmatrix}=\begin{bmatrix} M_{11} & \cdot \\ M_{21} & \cdot \end{bmatrix},$$

from which it follows that $L_{12}=M_{21}=0$. If this were true for every X, then the representations would be reducible, which is contrary to hypothesis. It follows that the rank r of Q is either 0 or f. If $r=0$, then $Q=0$; but if $r=f$, then Q is non-singular and possesses an inverse. In this case we have for every X

$$A_X=Q^{-1}B_XQ,$$

that is the two representations are equivalent.

In the preceding application of Schur's theorem we choose $A_X=U_X$, $B_X=W_X$, and take Q to be any one of the matrices $C_{r.v.}$ which does not vanish. Alternatively we can take $A_X=W_X$, $B_X=U_X$ and take Q to be any non-vanishing $C_{.s.v}'$. The theorem then tells us that the natural and semi-normal representations

$$\tau : W_\tau, \quad \tau : U_\tau$$

are equivalent.

The matrices $C_{r.v.}$ and $C_{.s.v}$ may be regarded as submatrices of the matrix of the transformation (31.1), the rows and columns of which are specified by double suffixes. We shall call this transformation matrix C. The element in the i,jth row and k,lth column is $c_{ij,kl}$ and we may suppose that the rows and columns are arranged in the order

$$1, 1\,;\ \ldots\,;\ 1, f\,;\ 2, 1\,;\ \ldots\,;\ 2, f\,;\ \ldots\,;\ f, 1\,;\ \ldots\,;\ f, f.$$

* T. A., p. 23.

Again, by (29.6), the coefficient of e_{ff} in g_{ff} is the coefficient of e_{ff} in $(1/\theta)e_{ff}E_{ff}e_{ff}$, which by (17.5) is +1. This shows us that the matrices $C_{f.f.}$ and $C_{.f.f}$ are non-zero and therefore, by Schur's theorem, non-singular. Now if $C_{r.v.}$ be also non-singular, we have

$$C_{r.v.}U_X C_{r.v.}^{-1} = W = C_{f.f.}U_X C_{f.f.}^{-1},$$

and so

$$C_{f.f.}^{-1}C_{r.v.}U_X = U_X C_{f.f.}^{-1}C_{r.v.}$$

for every matrix U_X. It follows from theorem 13 (§ 19) that $C_{f.f.}^{-1}C_{r.v.}$ is a scalar multiple of the unit matrix; that is,

$$C_{r.v.} = \gamma_{rv}C_{f.f.},$$

where γ_{rv} is numerical. This equation still holds when $C_{r.v.}$ is zero, for then we can take $\gamma_{rv}=0$. Obviously $\gamma_{ff}=1$.

It follows from this last equation that C is just the direct product of the matrix $[\gamma_{rv}]$ with $C_{f.f.}$ and that since $c_{ff,ff}=+1$ the matrix $[\gamma_{rv}]$ is identical to the matrix $C_{.f.f}$. Hence, using the notation $\times\cdot$ to denote the direct product,

$$C = C_{.f.f}\times\cdot\, C_{f.f.}.$$

Lastly, since we can write (31.4) in the form

$$W_X C_{.s.v}^{-1'} = C_{.s.v}^{-1'}U_X,$$

an argument similar to the above will show that

$$C_{.f.f}^{-1'} = kC_{f.f.},$$

where k is a numerical factor to be determined. Hence

$$I = kC_{.f.f}'C_{f.f.},$$

where I denotes the unit matrix. Equating the f,fth elements on both sides of this matrix equation, we find, using (31.2), that

$$1 = k\sum_t c_{ff,tf}c_{ff,ft} = k\delta_{ff}c_{ff,ff} = k.$$

We have therefore proved that

$$C_{.f.f}^{-1'} = C_{f.f.};$$

hence if we call this matrix Γ, then the matrix C of the transformation (31.1) is given by

$$C = \Gamma^{-1'}\times\cdot\,\Gamma,$$

where Γ is such that

$$\Gamma^{-1}W_X\Gamma = U_X$$

for every substitutional expression X.

CHAPTER V

GROUP CHARACTERS

Argument. Corresponding to any partition α of n there is a certain substitutional expression T^α. These expressions can be defined in many ways amongst which we may mention the following:

$$T^\alpha \equiv \sum_r e^\alpha_{rr} = \sum_r o^\alpha_{rr} = \sum_r g^\alpha_{rr}.$$

They satisfy the relations

$$T^\alpha T^\beta = \delta^{\alpha\beta} T^\alpha,$$
$$\sum_\alpha T^\alpha = \epsilon.$$

Associated with every partition β of n there is also a class of permutations the sum of whose elements is denoted by C_β. Each expression C_β is a linear combination of the expressions T^α and vice versa. In fact,

$$T^\alpha = (1/\theta^\alpha) \sum_\beta \chi^\alpha_\beta C_\beta,$$

where χ^α_β is a component of the group character associated with the representation α. The well-known orthogonal relations between the different characters are established and formulae are proved by means of which the different simple characters of the symmetric group may be calculated.

§ 32. The Expressions C_α

The concept of a class of permutations has already been defined in § 5. It was there shown that all permutations of $\mathcal{S}_n$ which are expressible as products of independent cycles of orders $\alpha_1, \ldots, \alpha_k$ belong to the same class. If we include cycles of order 1 where necessary and arrange the α_i suitably, we can ensure that

$$\alpha_1 \geqslant \ldots \geqslant \alpha_k, \quad \alpha_1 + \ldots + \alpha_k = n.$$

Thus each class is associated with a partition α of n and we may well call this the class $\mathfrak{C}_\alpha$. For example, the permutation (1, 2, 9, 6) (3, 8, 5) (4) (7) of $\mathcal{S}_9$ belongs to the class $\mathfrak{C}_{[4,3,1^2]}$.

Hitherto we have used an upper suffix α to indicate that the symbol concerned is associated with the shape α. We shall adhere to this practice but we now introduce lower suffixes α to indicate that the symbols under consideration are associated with the class $\mathfrak{C}_\alpha$.

We shall denote by h_α the number of distinct permutations belonging to $\mathfrak{C}_\alpha$. It was first shown by Cauchy that if $\alpha = [l^{v_l}, \ldots, 2^{v_2}, 1^{v_1}]$, then

$$h_\alpha = \frac{n!}{l^{v_l} \ldots 2^{v_2} 1^{v_1} v_l! \ldots v_2!\, v_1!}.$$

The proof of this formula is merely an exercise in permutations and combinations and will not be given here.

We shall further denote by C_α the sum of the h_α distinct permutations of $\mathfrak{C}_\alpha$. If τ belongs to $\mathfrak{C}_\alpha$, then by definition so does $\sigma\tau\sigma^{-1}$ for any σ. The permutations $\sigma\tau\sigma^{-1}$ are not however all distinct. Let $\epsilon, \sigma_2, \ldots \sigma_r$ be all the permutations σ for which

$$\sigma\tau\sigma^{-1} = \tau.$$

It is easily verified that such permutations form a subgroup of $\mathfrak{S}_n$ of order r. It is therefore possible to find permutations $\epsilon, \rho_2, \ldots, \rho_{n!/r}$ such that the $n!$ permutations $\rho_i\sigma_j (i = 1, \ldots, n!/r\,;\ j = 1, \ldots, r)$ are all distinct and thus comprise all the permutations of $\mathfrak{S}_n$. Now

$$(\rho_i\sigma_j)\tau(\rho_i\sigma_j)^{-1} = \rho_i\sigma_j\tau\sigma_j^{-1}\rho_i^{-1} = \rho_i\tau\rho_i^{-1}.$$

Hence in the summation

$$\sum_\sigma \sigma\tau\sigma^{-1},$$

where σ ranges over all the permutations of $\mathfrak{S}_n$, each permutation $\rho_i\tau\rho_i^{-1}$ $(i = 1, \ldots, n!/r)$ occurs exactly r times. Further, these permutation $\rho_i\tau\rho_i^{-1}$ are all distinct, for if

$$\rho_i\tau\rho_i^{-1} = \rho_j\tau\rho_j^{-1},$$

then

$$(\rho_j^{-1}\rho_i)\tau(\rho_j^{-1}\rho_i)^{-1} = \tau$$

and so

$$\rho_j^{-1}\rho_i = \sigma_k$$

say. That is,

$$\rho_i\epsilon = \rho_i = \rho_j\sigma_k,$$

which contradicts our assertion that the permutations $\rho_i\sigma_j$ are all distinct. This means that $n!/r = h_\alpha$ and that $\sum_i \rho_i\tau\rho_i^{-1} = C_\alpha$. Hence

$$\sum_\sigma \sigma\tau\sigma^{-1} = (n!/h_\alpha)C_\alpha,$$

provided τ belongs to the class $\mathfrak{C}_\alpha$.

Using the familiar property of groups that $\rho\sum_\sigma \sigma = \sum_\sigma \sigma$, where ρ is any permutation, we see that

$$\rho C_\alpha \rho^{-1} = h_\alpha/n! \sum_\sigma (\rho\sigma)\tau(\rho\sigma)^{-1} = h_\alpha/n! \sum_\sigma \sigma\tau\sigma^{-1} = C_\alpha.$$

Thus

$$\rho C_\alpha = C_\alpha \rho,$$

which shows that C_α commutes with every permutation and therefore with every substitutional expression.

Since no permutation can belong to more than one class, the linear independence of the expressions C_α follows at once from the linear independence of the permutations of $\mathfrak{S}_n$.

§ 33. The Expressions T^α

From equations (27.2) and (29.7) it is clear that

$$\sum_r e^\alpha_{rr} = \sum_r o^\alpha_{rr} = \sum_r g^\alpha_{rr}.$$

As this expression plays an important rôle in this chapter we shall give it a special symbol and denote it by T^α. Indeed, since the semi-normal matrix for T^α is the unit matrix, the corresponding matrix for any equivalent representation is of the form HIH^{-1}, that is, again the unit matrix. Thus if the units of any representation equivalent with the semi-normal one be a^α_{rs}, then

$$T^\alpha = \sum_r a^\alpha_{rr}.$$

Before we proceed further we shall derive other expressions for T^α. From equations (18.1) and (18.7) it follows that

$$\begin{aligned}
\sum_\sigma \sigma e^\alpha_{rs}\sigma^{-1} &= \sum_\sigma \sum_{t,v,\beta} u^\beta_{tv\sigma} e^\beta_{tv} e^\alpha_{rs} \sum_{p,q,\gamma} u^\gamma_{pq\sigma^{-1}} e^\gamma_{pq} \\
&= \sum_{\sigma,t,q} u^\alpha_{tr\sigma} u^\alpha_{sq\sigma^{-1}} e^\alpha_{tq} \\
&= \sum_{t,q} \delta_{tq}\delta_{rs}\theta^\alpha e^\alpha_{tq} \\
&= \sum_q \delta_{rs}\theta^\alpha e^\alpha_{qq} \\
&= \delta_{rs}\theta^\alpha T^\alpha.
\end{aligned}$$

Similarly,

$$\sum_\sigma \sigma o^\alpha_{rs}\sigma^{-1} = \sum_\sigma \sigma g^\alpha_{rs}\sigma^{-1} = \delta_{rs}\theta^\alpha T^\alpha.$$

It follows that for any value of r

$$T^\alpha = (1/\theta^\alpha)\sum_\sigma \sigma e^\alpha_{rr}\sigma^{-1} = (1/\theta^\alpha)\sum_\sigma \sigma o^\alpha_{rr}\sigma^{-1} = (1/\theta^\alpha)\sum_\sigma \sigma g^\alpha_{rr}\sigma^{-1}.$$

Since in particular we found in equation (29.6) that

$$g^\alpha_{ff} = (1/\theta^\alpha)E^\alpha_{ff},$$

it is easy to see that

$$T^\alpha = (1/\theta^\alpha)^2\sum_\sigma \sigma E^\alpha_{ff}\sigma^{-1} = (1/\theta^\alpha)^2\sum_\sigma \sigma\sigma^\alpha_{fr}E^\alpha_{rr}\sigma^\alpha_{rf}\sigma^{-1}.$$

But it is an elementary property of groups that $\sigma\sigma^\alpha_{fr}$ runs through all the elements of the group when σ does so. In consequence of this

$$T^\alpha = (1/\theta^\alpha)^2\sum_\sigma \sigma E^\alpha_{rr}\sigma^{-1},$$

where the suffix r is arbitrary. Further, the expression $\sum_\sigma \sigma E^\alpha_{rr}\sigma^{-1}$ is merely $\sum_r E^\alpha_{rr}$, where r is summed over all tableaux of shape α whether standard or not. Thus

$$T^\alpha = (1/\theta^\alpha)^2 \sum_r E^\alpha_{rr}.$$

This last equation amounts to Young's original definition of the expression T^α apart from the numerical factor $(1/\theta^\alpha)^2$ which he at first omitted.*

Again, if τ be any permutation,

$$\tau T^\alpha \tau^{-1} = (1/\theta^\alpha)^2 \sum_\sigma \tau\sigma E^\alpha_{rr}\sigma^{-1}\tau^{-1} = (1/\theta^\alpha)^2 \sum_\sigma \sigma E^\alpha_{rr}\sigma^{-1} = T^\alpha.$$

Thus for any τ

$$\tau T^\alpha = T^\alpha \tau;$$

that is to say, T^α commutes with every permutation and therefore with any substitutional expression whatever.†

We are now in a position to proceed to the long-delayed evaluation of θ^α. Summing the relation

$$\theta^\alpha T^\alpha = \sum_\sigma \sigma e^\alpha_{rr}\sigma^{-1}$$

over the suffix r, we obtain

$$f^\alpha \theta^\alpha T^\alpha = \sum_\sigma \sigma T^\alpha \sigma^{-1} = n!\, T^\alpha.$$

Since the coefficient of ϵ in T^α is f^α/θ^α, no T^α vanishes and therefore ‡

$$\theta^\alpha = \frac{n!}{f^\alpha}. \qquad \ldots(33.1)$$

Substituting from (16.1) in (33.1), we obtain

$$\theta^\alpha = \frac{\prod_r x_r!}{\prod_{r<s\leqslant k}(x_r - x_s)},$$

which may be written

$$\theta^\alpha = \prod_r \left\{\frac{x_r!}{\prod_{r<s\leqslant k}(x_r - x_s)}\right\}.$$

Now if $r<s$, then $x_r > x_s$ since $\alpha_r \geqslant \alpha_s$. It follows that each factor

$$\frac{x_r!}{\prod_{r<s\leqslant k}(x_r - x_s)}$$

is a positive integer and so θ^α is itself a positive integer. Since f^α is also a positive integer, both θ^α and f^α are integral factors of $n!$

It is an immediate consequence of the result

$$\sum_{\alpha, r} e^\alpha_{rr} = \epsilon$$

* Young I, p. 133. † Young II, p. 369. ‡ Young II, p. 367.

proved in § 17 that

$$\sum_{\alpha} T^{\alpha} = \epsilon. \qquad \ldots(33.2)$$

This is a formula by which Young was greatly fascinated. Again, from (17.4) it follows at once that

$$\sum_{r} e^{\alpha}_{rr} \sum_{v} e^{\beta}_{vv} = \delta^{\alpha\beta} \sum_{r,v} \delta_{rv} e^{\alpha}_{rv} = \delta^{\alpha\beta} \sum_{r} e^{\alpha}_{rr},$$

that is,*

$$T^{\alpha} T^{\beta} = \delta^{\alpha\beta} T^{\alpha}. \qquad \ldots(33.3)$$

The expressions T^{α} are linearly independent. For suppose that a non-trivial relation $\sum_{\alpha} \lambda^{\alpha} T^{\alpha} = 0$ exists; then in virtue of (33.3) we should have

$$0 = T^{\beta} \sum_{\alpha} \lambda^{\alpha} T^{\alpha} = \lambda^{\beta} T^{\beta},$$

indicating that, contrary to hypothesis, each λ^{β} vanishes.

§ 34. The Relations between the Expressions T^{α} and C_{α}

Since, as was shown in § 33,

$$n!\, T^{\alpha} = \sum_{\sigma} \sigma T^{\alpha} \sigma^{-1}$$

and since

$$\sum_{\sigma} \sigma \tau \sigma^{-1} = (n!/h_{\beta}) C_{\beta}$$

if τ belongs to $\mathfrak{C}_{\beta}$, it is patent that T^{α} must be a linear combination of the expressions C_{β}. We may express this by writing †

$$T^{\alpha} = (1/\theta^{\alpha}) \sum_{\beta} \chi^{\alpha}_{\beta} C_{\beta}, \qquad \ldots(34.1)$$

where the coefficients χ^{α}_{β} are numerical. The reason for including the factor $(1/\theta^{\alpha})$ will appear in § 35 where it will be shown that χ^{α}_{β} is a component of a simple group character. Since there is one T^{α} and one C_{α} associated with each partition α, there is the same number of each type of expression. Further, each set is linearly independent. We can therefore solve the equations (34.1) for the C_{β} in terms of the T^{α}. Let us write this solution in the form ‡

$$C_{\beta} = \sum_{\alpha} \psi^{\alpha}_{\beta} T^{\alpha}. \qquad \ldots(34.2)$$

Equation (33.2) is a particular case of the last equation. In fact

$$\psi^{\alpha}_{[1^n]} = 1. \qquad \ldots(34.3)$$

By combining equations (34.1) and (34.2) we find that

$$T^{\alpha} = (1/\theta^{\alpha}) \sum_{\beta} \chi^{\alpha}_{\beta} \sum_{\gamma} \psi^{\gamma}_{\beta} T^{\gamma}$$

and

$$C_{\beta} = \sum_{\alpha} \psi^{\alpha}_{\beta} (1/\theta^{\alpha}) \sum_{\gamma} \chi^{\alpha}_{\gamma} C_{\gamma}.$$

* Young I, p. 139. † Young I, p. 136; Young IV, p. 256.
‡ Young I, p. 137.

Since both the sets T^α and C_β are linearly independent, we conclude that

$$(1/\theta^\alpha)\sum_\beta \chi^\alpha_\beta \psi^\gamma_\beta = \delta^{\alpha\gamma} \qquad \ldots(34.4)$$

and

$$\sum_\alpha (1/\theta^\alpha)\psi^\alpha_\beta \chi^\alpha_\gamma = \delta_{\beta\gamma}. \qquad \ldots(34.5)$$

Since the T^α form a linearly independent set, the matrix $[\chi^\alpha_\beta]$ is non-singular and so the above equations can be solved for the ψ^α_β in terms of the χ^α_β. This solution will be exhibited in the next section.

§ 35. The Group Characters of $\mathfrak{S}_n$

The *trace* of a square matrix is the sum of its diagonal elements. It is an established fact that the traces of the matrices A and HAH^{-1} are the same, where A and H are square and H is non-singular. Now if τ_1 and τ_2 are two permutations of the same class, then $\tau_1 = \sigma\tau_2\sigma^{-1}$ for some σ and hence, with the notation of § 19,

$$U^\alpha_{\tau_1} = U^\alpha_\sigma U^\alpha_{\tau_2} U^\alpha_{\sigma^{-1}}.$$

Further, since the product $U^\alpha_\sigma U^\alpha_{\sigma^{-1}}$ is the unit matrix, U^α_σ is non-singular. It follows that the traces of $U^\alpha_{\tau_1}$ and $U^\alpha_{\tau_2}$ are the same provided τ_1 and τ_2 belong to the same class $\mathfrak{C}_\beta$. In other words, for any representation α the trace of any element of $\mathfrak{C}_\beta$ is the same. This trace is called the *β-component* of the character associated with the representation α. We shall show presently that this component is precisely χ^α_β. The *character* of the representation α is the set of these components. It has one component χ^α_β for each class $\mathfrak{C}_\beta$.

The character of an irreducible representation is said to be *simple*. As all the representations under consideration have been shown to be irreducible, so all their characters are simple.

If τ belong to $\mathfrak{C}_\beta$, then by definition the β-component of the character associated with the representation α is

$$\sum_r u^\alpha_{rr\tau}.$$

But $u^\alpha_{rr\tau}$ is, as has been shown in (18.4), the coefficient of τ^{-1} in $\theta^\alpha e^\alpha_{rr}$. The required component is therefore the coefficient of τ^{-1} in $\theta^\alpha \sum_r e^\alpha_{rr}$, that is, in $\theta^\alpha T^\alpha$, or, in $\sum_\beta \chi^\alpha_\beta C_\beta$. But since in the case of $\mathfrak{S}_n$ the permutations τ and τ^{-1} belong to the same class, this coefficient is simply χ^α_β. It follows that χ^α_β is the component, associated with the class $\mathfrak{C}_\beta$, of the representation α.

The equation

$$\sum_\tau u^\alpha_{rr\tau^{-1}} u^\beta_{ss\tau} = \delta^{\alpha\beta}\delta_{rs}\theta^\alpha$$

is a particular case of (18.7). By summing it over r and s, we find that

$$\sum_\tau \chi^\alpha_{\tau^{-1}}\chi^\beta_\tau = \delta^{\alpha\beta} f^\alpha \theta^\alpha = \delta^{\alpha\beta} n!,$$

where χ^{β}_{τ} stands for χ^{β}_{γ} if τ and therefore τ^{-1} belong to $\mathfrak{C}_{\gamma}$. Since there are h_{γ} elements in the class $\mathfrak{C}_{\gamma}$, the above equation may be written

$$\sum_{\gamma} h_{\gamma}\chi^{\alpha}_{\gamma}\chi^{\beta}_{\gamma}=\delta^{\alpha\beta}n!. \qquad \ldots(35.1)$$

This relation displays the unique solution of (34.4). Thus we have

$$\psi^{\beta}_{\gamma}=(h_{\gamma}/f^{\beta})\chi^{\beta}_{\gamma}.$$

Substituting this result in (34.2) and (34.5), we immediately obtain

$$C_{\beta}=h_{\beta}\sum_{\alpha}(1/f^{\alpha})\chi^{\alpha}_{\beta}T^{\alpha} \qquad \ldots(35.2)$$

and

$$h_{\beta}\sum_{\alpha}\chi^{\alpha}_{\beta}\chi^{\alpha}_{\gamma}=\delta_{\beta\gamma}n!. \qquad \ldots(35.3)$$

From the equation (34.3) we find that

$$\chi^{\alpha}_{[1^n]}=f^{\alpha},$$

whence, substituting in (35.3), we obtain the relation

$$\sum_{\alpha} f^{\alpha}\chi^{\alpha}_{\gamma}=\delta_{[1^n]\gamma}n!. \qquad \ldots(35.4)$$

Again, $\chi^{[n]}_{\beta}$ is the coefficient of any permutation τ belonging to $\mathfrak{C}_{\beta}$ in $\theta^{[n]}T^{[n]}$, or, since there is only one standard tableau of shape $[n]$, in $\theta^{[n]}e^{[n]}_{11}$. By (18.4) this is seen to be $u^{[n]}_{11\tau}$, and this in turn by (23.2) is just $+1$. Hence for all β

$$\chi^{[n]}_{\beta}=+1.$$

Similarly,

$$\chi^{[1^n]}_{\beta}=\pm 1,$$

according as $\mathfrak{C}_{\beta}$ consists of even or odd permutations. Now if we put $\beta=[n]$ in equation (35.1) and substitute for $\chi^{[n]}_{\gamma}$, we get

$$\sum_{\gamma} h_{\gamma}\chi^{\alpha}_{\gamma}=\delta^{\alpha[n]}n!.$$

Similarly,

$$\sum_{\gamma}\zeta_{\gamma}h_{\gamma}\chi^{\alpha}_{\gamma}=\delta^{\alpha[1^n]}n!,$$

where ζ_{γ} is $+1$ or -1 according as $\mathfrak{C}_{\gamma}$ consists of even or odd permutations.

With the help of equations (35.2) and (34.1) we can now construct a multiplication table for the expressions C_{β}. In fact,

$$\begin{aligned}C_{\beta}C_{\gamma}&=h_{\beta}h_{\gamma}\sum_{\alpha}(1/f^{\alpha})\chi^{\alpha}_{\beta}T^{\alpha}\sum_{\delta}(1/f^{\delta})\chi^{\delta}_{\gamma}T^{\delta}\\&=h_{\beta}h_{\gamma}\sum_{\alpha}(1/f^{\alpha})^{2}\chi^{\alpha}_{\beta}\chi^{\alpha}_{\gamma}T^{\alpha}\\&=h_{\beta}h_{\gamma}\sum_{\alpha}(1/f^{\alpha})^{2}\chi^{\alpha}_{\beta}\chi^{\alpha}_{\gamma}(f^{\alpha}/n!)\sum_{\delta}\chi^{\alpha}_{\delta}C_{\delta}\\&=\sum_{\delta}c_{\beta\gamma,\delta}C_{\delta},\end{aligned}$$

where

$$c_{\beta\gamma,\delta}=\frac{h_{\beta}h_{\gamma}}{n!}\sum_{\alpha}\frac{\chi^{\alpha}_{\beta}\chi^{\alpha}_{\gamma}\chi^{\alpha}_{\delta}}{f^{\alpha}}=c_{\gamma\beta,\delta}.$$

§ 36. The Evaluation of the Group Characters

The formula

$$T^\alpha = (1/\theta^\alpha)^2 \sum_\sigma \sigma E_{rr}^\alpha \sigma^{-1}$$

may be used to evaluate the character components χ_β^α. We have seen in theorem 7 (§ 11) that the coefficient of any permutation in E_{rr}^α takes one of the values $+1$, -1, 0. Suppose that there are a_β^α permutations of the class $\mathfrak{C}_\beta$ which appear in E_{rr}^α with the coefficient $+1$ and b_β^α permutations with the coefficient -1. For any value of σ the expression $\sigma E_{rr}^\alpha \sigma^{-1}$ must also have a_β^α permutations of class $\mathfrak{C}_\beta$ with coefficient $+1$ and b_β^α with coefficient -1. It follows that the sum of the coefficients in T^α of the permutations of class $\mathfrak{C}_\beta$ is $(1/\theta^\alpha)^2 n!(a_\beta^\alpha - b_\beta^\alpha)$. On the other hand, equation (34.1) shows that this sum is precisely $\chi_\beta^\alpha h_\beta/\theta^\alpha$. Hence, since $n!/\theta^\alpha = f^\alpha$,

$$\chi_\beta^\alpha = (f^\alpha/h_\beta)(a_\beta^\alpha - b_\beta^\alpha).$$

This result can be expressed slightly differently in the following manner.

THEOREM 21. *The algebraic sum of the coefficients in any* E_{rr}^α *of all the permutations belonging to* $\mathfrak{C}_\beta$ *is*

$$\frac{h_\beta \chi_\beta^\alpha}{f^\alpha}.$$

For example, if S be the tableau

1	**2**
3	**4**

of shape $[2^2]$, then $f^{[2^2]} = 2$ and

$$E = (\epsilon + (1,2) + (3,4) - (1,3) - (2,4)) + (-(3,2,1) - (4,3,1) - (1,2,4) - (2,3,4))$$
$$+ ((1,2)(3,4) + (1,3)(2,4) + (1,4)(2,3)) + ((1,3,2,4) + (3,1,4,2) - (4,3,2,1) - (1,2,3,4)).$$

The various components $\chi_\beta^{[2^2]}$ are now obtained in the following tabulation:

$\beta =$	$[1^4]$	$[2, 1^2]$	$[3, 1]$	$[2^2]$	$[4]$
$h_\beta =$	1	6	8	3	6
$h_\beta \chi_\beta^{[2^2]}/f^{[2^2]} =$	1	0	-4	3	0
$\chi_\beta^\alpha =$	2	0	-1	2	0

In a calculation such as the preceding one, the expression for E can either be obtained by the direct multiplication of the expressions for P and N or else by the schematic method consequent on theorem 7 (§ 11). The latter alternative is usually the less arduous especially if only one component of the character is required.

Alternative methods for the evaluation of χ_β^α are available when n is small. The component χ_β^α is the coefficient of C_β, and therefore the coefficient of any permutation belonging to $\mathfrak{C}_\beta$, in $\theta^\alpha T^\alpha$. For example, the coefficients of the transposition **(1, 2)** in $\theta^{[3,1]}g_{11}^{[3,1]}$, $\theta^{[3,1]}g_{22}^{[3,1]}$, $\theta^{[3,1]}g_{33}^{[3,1]}$ are respectively $+1, +1, -1$. Hence

$$\chi_{[2,1^2]}^{[3,1]} = +1+1-1 = +1.$$

The foregoing methods evidently have their limitations on account of the enormous labour involved when n attains values greater than 4 or 5. Formulae are, however, provided by which the different characters of $\mathfrak{S}_n$ can be derived from those of $\mathfrak{S}_{n-1}, \mathfrak{S}_{n-2}, \ldots$ without undue manipulation. These will be obtained in the next section.

Tables of character components χ_β^α for symmetric groups of different orders are listed elsewhere.* We include here as specimens those for $\mathfrak{S}_3$ and $\mathfrak{S}_4$.

$\mathfrak{S}_2$:

α	[3]	[2, 1]	$[1^3]$	β
[3]	1	1	1	1
[2, 1]	−1	0	2	2
$[1^3]$	1	−1	1	1
h_β	2	3	1	f^α

$\mathfrak{S}_4$:

α	[4]	[3, 1]	$[2^2]$	$[2, 1^2]$	$[1^4]$	β
[4]	1	1	1	1	1	1
[3, 1]	−1	0	−1	1	3	3
$[2^2]$	0	−1	2	0	2	2
$[2, 1^2]$	1	0	−1	−1	3	3
$[1^4]$	−1	1	1	−1	1	1
h_β	6	8	3	6	1	f^α

§ 37. Reduction Formulae

Let $\beta = [\beta_1, \ldots, \beta_{h-1}, 1]$ be a partition of n and let us write $\beta' = [\beta_1, \ldots, \beta_{h-1}]$. That is to say, the elements of $\mathfrak{C}_\beta$ have at least one cycle of order unity and there are therefore permutations of $\mathfrak{C}_\beta$ which do not affect the last letter n. Further, these permutations can also be regarded as belonging to the class $\mathfrak{C}_{\beta'}$ of $\mathfrak{S}_{n-1}$. It follows from theorem 15 (§ 20) that if τ be any one of these permutations, then

$$U_\tau^\alpha = U_\tau^{[\alpha_k -]} \dotplus \ldots \dotplus U_\tau^{[\alpha_1 -]}.$$

* Littlewood, Appendix.

Since χ^α_β is the sum of the diagonal elements of U^α_τ and $\chi^{[\alpha_i-]}_{\beta'}$ is the sum of the diagonal elements of $U^{[\alpha_i-]}_\tau$, it follows that

$$\chi^\alpha_\beta = \chi^{[\alpha_k-]}_{\beta'} + \ldots + \chi^{[\alpha_1-]}_{\beta'}. \qquad \ldots(37.1)$$

For example,

$$\chi^{[4,3^2,1]}_{[7,3,1]} = \chi^{[4,3^2]}_{[7,3]} + \chi^{[4,3,2,1]}_{[7,3]} + \chi^{[3^2,1]}_{[7,3]}.$$

The formula (37.1) gives the character component χ^α_β of $\mathfrak{S}_n$ in terms of the character components of $\mathfrak{S}_{n-1}$, provided that the class $\mathfrak{C}_\beta$ contains at least one cycle of order unity. The corresponding formula when $\mathfrak{C}_\beta$ contains no cycle of order unity is a generalisation of the preceding one. It is not really difficult to prove, but the proof which follows involves a somewhat lengthy description of the arrangements of the letters in the different tableaux under consideration. The reader may, however, be already familiar with shorter proofs developed without the aid of Substitutional Analysis.*

Let $\beta = [\beta_1, \ldots, \beta_h]$ be a partition of n and let us now write $\beta' = [\beta_1, \ldots, \beta_{h-1}]$. Suppose that the last β_h letters are the letters $\mathbf{m+1}, \mathbf{m+2}, \ldots, \mathbf{n}$. There is therefore at least one permutation of $\mathfrak{C}_\beta$ which is of the form $\sigma\tau$, where σ is of class $\mathfrak{C}_{\beta'}$ and involves the letters $\mathbf{1}, \mathbf{2}, \ldots, \mathbf{m}$ only and where τ is the cycle

$$(\mathbf{m+1}, \mathbf{m+2}, \ldots, \mathbf{n}).$$

By making $n-m$ repeated applications of theorem 15 (§ 20) the matrix U^α_σ will be seen to take the form of a direct sum of submatrices $U^{[\alpha_{i_1}-, \ldots, \alpha_{i_{n-m}}-]}_\sigma$. The shapes $[\alpha_{i_1}-, \ldots, \alpha_{i_{n-m}}-]$ involved are those obtained from α by removing first the last space of the i_1th row, next the last space of the i_2th row and so on, assuming that at each stage the new tableau is a recognised one. For example, if $n=8$, $m=6$,

$$\begin{aligned} U^{[3^2,2]}_\sigma &= U^{[\alpha_3-,\alpha_3-]}_\sigma + U^{[\alpha_3-,\alpha_2-]}_\sigma + U^{[\alpha_2-,\alpha_3-]}_\sigma + U^{[\alpha_2-,\alpha_1-]}_\sigma \\ &= U^{[3^2]}_\sigma + U^{[3,2,1]}_\sigma + U^{[3,2,1]}_\sigma + U^{[2^3]}_\sigma. \end{aligned}$$

As illustrated in this example, amongst these submatrices there may be repetitions. This is because the same shape α' of m spaces may arise in different ways in the form $[\alpha_{i_1}-, \ldots, \alpha_{i_{n-m}}-]$.

Since the cycle τ commutes with every permutation belonging to $\mathfrak{S}_m$, it follows from theorem 13 (§ 19), by an argument similar to that used in theorem 17 (§ 22), that when the matrix U^α_τ is partitioned in the same way as U^α_σ, each submatrix on the leading diagonal is scalar. The submatrices on the leading diagonal of the product $U^\alpha_{\sigma\tau}$ are therefore each of the form

$$c^{[\alpha_{i_1}-, \ldots, \alpha_{i_{n-m}}-]} U^{[\alpha_{i_1}-, \ldots, \alpha_{i_{n-m}}-]},$$

where the coefficients $c^{[\alpha_{i_1}-, \ldots, \alpha_{i_{n-m}}-]}$ are numerical. It follows that

$$\chi^\alpha_\beta = \Sigma c^{[\alpha_{i_1}-, \ldots, \alpha_{i_{n-m}}-]} \chi^{[\alpha_{i_1}-, \ldots, \alpha_{i_{n-m}}-]}_{\beta'},$$

since χ^α_β is the sum of the diagonal elements of $U^\alpha_{\sigma\tau}$.

* Murnaghan, p. 460; Nakayama, p. 182.

We have seen that a particular shape α' of m letters may arise in different ways in the form $[\alpha_{i_1}-, \ldots \alpha_{i_{n-m}}-]$. For this reason there may be several terms in the above summation for χ^{α}_{β} which contain the same factor $\chi^{\alpha'}_{\beta'}$, but it does not follow that the coefficient $c^{[\alpha_{i_1}-, \ldots, \alpha_{i_{n-m}}-]}$ is the same in each of these terms. We can, however, write the last equation in the form

$$\chi^{\alpha}_{\beta} = \sum_{\alpha'} d^{\alpha'} \chi^{\alpha'}_{\beta'}, \qquad \ldots (37.2)$$

where $d^{\alpha'}$ is the sum of these coefficients $c^{[\alpha_{i_1}-, \ldots, \alpha_{i_{n-m}}-]}$ taken over all shapes $[\alpha_{i_1}-, \ldots, \alpha_{i_{n-m}}-]$ which are of shape α'.

Now, expressing τ as a product of transpositions,

$$\tau = (\mathbf{m+1}, \mathbf{m+2}, \ldots, \mathbf{n}) = (\mathbf{m+1}, \mathbf{m+2})(\mathbf{m+2}, \mathbf{m+3}) \ldots (\mathbf{n-1}, \mathbf{n}),$$

it will be seen that any coefficient $c^{[\alpha_{i_1}-, \ldots, \alpha_{i_{n-m}}-]}$ can be evaluated by means of the deduction at the end of § 23. The fact that the submatrix $U^{[\alpha_{i_1}-, \ldots, \alpha_{i_{n-m}}-]}_{\tau}$ is scalar is a consequence of the fact that the coefficient $c^{[\alpha_{i_1}-, \ldots, \alpha_{i_{n-m}}-]}$ is so determined, solely by the positions of the letters $\mathbf{m+1}, \ldots, \mathbf{n}$. All the tableaux specifying the rows and columns of this submatrix have the letters $\mathbf{m+1}, \ldots, \mathbf{n}$ similarly placed.

When the letters $\mathbf{1}, \ldots, \mathbf{m}$ are removed from a standard tableau of shape α, there remains what Young called a *distorted tableau* * containing only the letters $\mathbf{m+1}, \ldots, \mathbf{n}$. This distorted tableau is said to be of *distorted shape* $\alpha - \alpha'$ if the m deleted letters constitute a tableau of shape α'. Each distorted tableau of shape $\alpha - \alpha'$ will thus give rise to one term in the summation

$$d^{\alpha'} = \sum c^{[\alpha_{i_1}-, \ldots, \alpha_{i_{n-m}}-]}.$$

Any distorted shape must either (case (i)) contain a square block of four spaces, or else (case (ii)) it does not do so. In case (ii) it must consist of one or more angular strips one space in width. In the illustrations below the distorted shapes are indicated by heavy lines, case (ii) having three strips of lengths 5, 1 and 3.

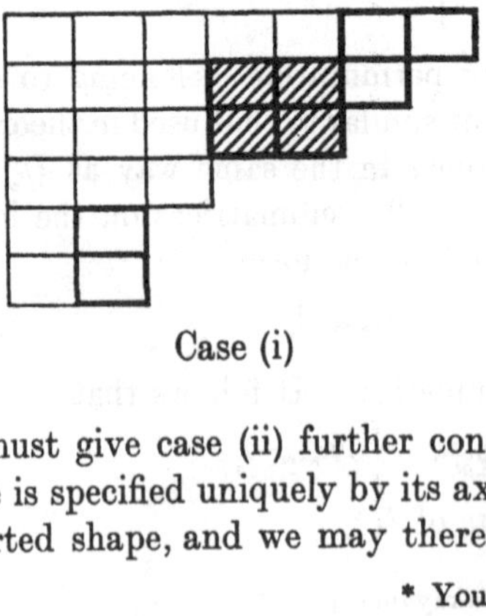

Case (i)

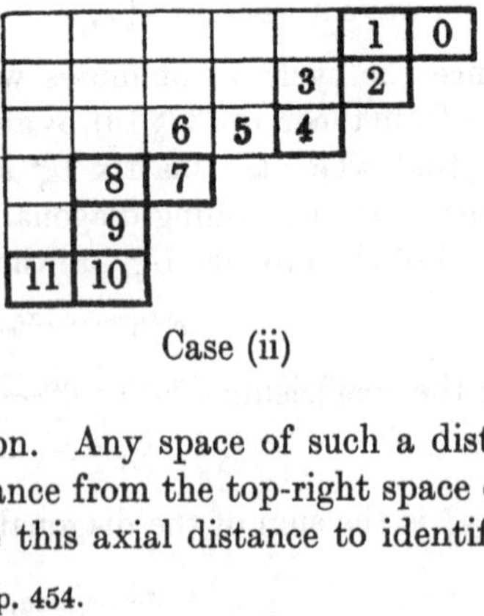

Case (ii)

We must give case (ii) further consideration. Any space of such a distorted shape is specified uniquely by its axial distance from the top-right space of the distorted shape, and we may therefore use this axial distance to identify the

* Young VIII, p. 454.

different spaces. This does not hold in case (i) where the top-left and bottom-right spaces of the square block are at the same axial distance from the top-right space of the distorted shape. Thus in the illustration of case (ii) the spaces of the distorted shape may be referred to as the spaces 0, 1, 2, 4, 7, 8, 9, 10, 11.

Again, in case (ii), we designate any position where the earliest letter $\mathbf{m+1}$ could appear as an *upper node* and any possible position of the last letter $\mathbf{n}$ which is not at the end of a strip we call a *lower node*. In our illustration the upper nodes are the spaces 1, 4, 8, 11. The only lower node is the space 10. In general, we may suppose that the upper nodes occur in the spaces $a_1, \ldots, a_u$ and the lower nodes in the spaces $b_1, \ldots, b_v$. It will be noticed that in any strip the number of upper nodes exceeds the number of lower nodes by unity. This fact will be required presently.

In the evaluation of $d^{\alpha'}$ we must consider distorted tableaux in which the earliest letter $\mathbf{m+1}$ appears in different upper nodes, but we shall first find the sum $q_i^{\alpha'}$ of the coefficients $c^{[\alpha_{i_1}-,\ldots,\alpha_{i_{n-m}}-]}$ summed over those distorted tableaux of shape $\alpha-\alpha'$ in which the letter $\mathbf{m+1}$ occurs in a fixed position a_i. When this has been done, $d^{\alpha'}$ will be given by

$$d^{\alpha'} = \sum_i q_i^{\alpha'}. \qquad \ldots(37.3)$$

Lemma. *In case* (i) $q_i^{\alpha'} = 0$; *in case* (ii)

$$q_i^{\alpha'} = \pm \frac{\prod_s (b_s - a_i)}{\prod_{r \neq i} (a_r - a_i)},$$

where the sign is to be taken as + or − according as there are an even or an odd number of vertical steps in the distorted shape.

Before we establish this lemma, we shall illustrate its content with reference to our illustration of case (ii). The number of vertical steps in the component strips of lengths 5, 1, 3 are respectively 2, 0, 1. Thus if a_i be the space 8, then

$$q_i^{\alpha'} = (-)^{2+0+1} \frac{(10-8)}{(1-8)(4-8)(11-8)} = -\frac{1}{42}.$$

Proof. We proceed by induction the basis of which are the cases $n-m=1, 2$. The truth of the lemma can be verified directly in these cases with the aid of the deduction at the end of § 23.

We consider first of all case (ii). In this case the sum $q_i^{\alpha'}$ is determined by tableaux all of which have $\mathbf{m+1}$ in the space a_i. In such tableaux the letters $\mathbf{m+2}, \ldots, \mathbf{n}$ must appear in a distorted shape $\alpha-\alpha_i'$, where α_i' is the shape obtained by adding the space a_i to the shape α'. If the space a_i is isolated, that is, if it forms by itself a strip of unit length, then the upper nodes of the shape $\alpha-\alpha_i'$ are the same as the upper nodes of the shape $\alpha-\alpha'$ with the exception of a_i, and the lower nodes are the same in both cases. If a_i is not isolated it may appear at the top end, in the middle, or at the lower end of a strip of the shape $\alpha-\alpha'$. In these cases the shape $\alpha-\alpha_i'$ will have, in addition to those of $\alpha-\alpha'$, upper nodes a_i+1, or a_i+1 and a_i-1, or a_i-1 in the respective cases mentioned. The lower nodes will be the same as before except when $a_i=b_j+1$, in which case the lower node b_j will not be a lower node of $\alpha-\alpha_i'$. The proofs in these different cases are essentially the same so we shall content ourselves with the case where a_i is isolated. The slight modifications required in the other cases can be supplied by the reader.

Of the tableaux of shape $\alpha-\alpha'$ which have **m + 1** in the space a_i there will be a certain number which also have **m + 2** in the space a_j. In view of the deduction at the end of § 23, these tableaux contribute a term $q_j^{\alpha_i'}/(a_i-a_j)$ to the sum $q_i^{\alpha'}$, for a_i-a_j is the axial distance from space a_j to space a_i. Thus,

$$q_i^{\alpha'}=\sum_{j\neq i}\frac{1}{(a_i-a_j)}q_j^{\alpha_i'}. \qquad \ldots(37.4)$$

Also, by the induction hypothesis

$$q_j^{\alpha_i'}=\pm\frac{\prod_s (b_s-a_j)}{\prod_{r\neq i,j}(a_r-a_j)},$$

whence, by partial fractions,

$$q_i^{\alpha'}=\pm\sum_{j\neq i}\frac{1}{(a_i-a_j)}\frac{\prod_s (b_s-a_j)}{\prod_{r\neq i,j}(a_r-a_j)}=\pm\frac{\prod_s (b_s-a_i)}{\prod_{r\neq i}(a_r-a_i)},$$

as required. It is to be observed that when a_i is isolated the shapes $\alpha-\alpha'$ and $\alpha-\alpha_i'$ have the same number of upward steps, from which we see that the $\pm$ sign just obtained is the same as that stated in our lemma. The modifications required in the above proof when a_i is not isolated are bound up with the adjustment of the $\pm$ sign by means of additional factors such as (a_i+1-a_i) or (a_i-1-a_i).

Turning now to case (i) in which the distorted shape $\alpha-\alpha'$ contains a square block, we find that we can still use the formula (37.4) if we understand a_i and a_j to mean the axial distances of **m + 1** and **m + 2** respectively from the top-right space of the distorted shape. By the induction hypothesis each $q_j^{\alpha_i'}$ vanishes with possibly one exception. The exception is the case where a strip yields a block of four letters when the letter **m + 1** is added to it. In such a case **m + 1** is in the top-left corner of the block, but there must also be a lower node b_j of the strip in the bottom-right corner such that $b_j=a_i$ numerically. The remaining $q_j^{\alpha_i'}$ contains the factor (b_j-a_i) and therefore vanishes also. Thus in case (i) $q_i^{\alpha'}=0$. This concludes the proof of the lemma.

Returning now to equation (37.3),

$$d^{\alpha'}=\sum_i q_i^{\alpha'},$$

it is clear that if the distorted shape $\alpha-\alpha'$ contains a square block, each $q_i^{\alpha'}$

vanishes and so $d^{\alpha'}=0$. If the shape $\alpha-\alpha'$ does not contain a square block, then from the preceding lemma

$$d^{\alpha'} = \pm \sum_i \frac{\prod_s (b_s - a_i)}{\prod_{r \neq i} (a_r - a_i)}.$$

If the distorted shape consists of several strips, the degree of the numerators is less than the degree of the denominators. In this case the right-hand side of the above equation vanishes by a well-known identity which is equivalent to Lagrange's interpolation formula. If the distorted shape consists of one continuous strip, then the degrees of the numerators and denominators are the same. In this case a corresponding identity yields

$$d^{\alpha'} = \pm 1,$$

the sign being determined as before. Substituting these results in equation (37.2) we come to the following conclusion.

Theorem 22. *With the notation of the present section*

$$\chi^{\alpha}_{\beta} = \sum_{\alpha'} d^{\alpha'} \chi^{\alpha'}_{\beta'},$$

where $d^{\alpha'}=0$ *unless* α' *is obtained from* α *by removing a continuous but possibly angular strip of* $n-m$ *spaces from its lower edge, in which case* $d^{\alpha'}=\pm 1$ *according as the strip contains an even or an odd number of vertical steps.*

In illustration, let us apply this theorem to the evaluation of $\chi^{[3^2,1]}_{[4,3]}$. If we remove 3 spaces from the shape $[3^2, 1]$, the appropriate shapes α' are $[3, 1]$, $[2^2]$, $[2, 1^2]$ and the corresponding distorted shapes are

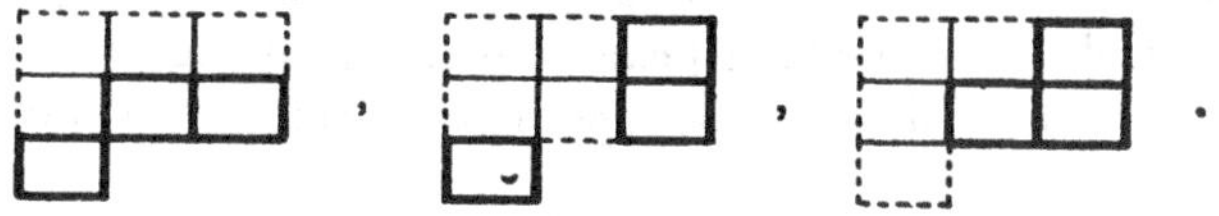

Of these, only the last is a continuous strip and it involves only one vertical step. Thus

$$\chi^{[3^2,1]}_{[4,3]} = -\chi^{[2,1^2]}_{[4]} = -1.$$

It has probably been observed by the reader that our proof of theorem 22 was not consequent on the cycle of order β_h being the smallest cycle in the elements of $\mathfrak{C}_\beta$. We can therefore evaluate $\chi^{[3^2,1]}_{[4,3]}$ alternatively by removing four spaces from the shape $[3^2, 1]$. In this case the appropriate values for α' are $[3]$, $[2, 1]$, $[1^3]$ and the distorted tableaux are

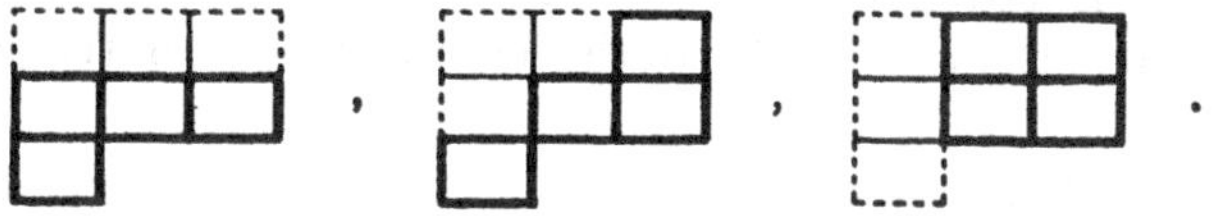

Of these, only the first is a continuous strip and it involves one vertical step. Hence

$$\chi^{[3^2,1]}_{[4,3]} = -\chi^{[3]}_{[3]} = -1.$$

§ 38. Littlewood's Theorem

The continuous strip of β_h spaces which when added to a shape α' yields a shape α is called by Littlewood a *regular application.* It is said to be *positive* or *negative* according as the number of vertical steps in it is even or odd. If we apply theorem 22 (§ 37) to each $\chi_{\beta'}^{\alpha'}$ in the formula

$$\chi_{\beta}^{\alpha}=\sum_{\alpha'} d^{\alpha'}\chi_{\beta'}^{\alpha'},$$

we obtain

$$\chi_{\beta}^{\alpha}=\sum_{\alpha'} d^{\alpha'}\sum_{\alpha''} d^{\alpha''}\chi_{\beta''}^{\alpha''},$$

where $\beta''=[\beta_1, \ldots, \beta_{h-2}]$, where in each second summation α'' ranges over such partitions of $n-\beta_h-\beta_{h-1}$ as yield α' when a regular application of β_{h-1} spaces is made to α'', and where $d^{\alpha''}$ is ± 1 according as the total number of vertical steps in this application are even or odd. Repeated application of theorem 22 will eventually exhaust all the h cycles of class $\mathfrak{C}_\beta$ and yield the following result.

THEOREM 23 (Littlewood's Theorem *). *The character component χ_{β}^{α} is given by the formula*

$$\chi_{\beta}^{\alpha}=\sum_{i} d_i,$$

where there is one term d_i for each way in which the shape α can be built up by first making a regular application of β_1 spaces, secondly a regular application of β_2 spaces, . . ., and lastly a regular application of β_h spaces, and where $d_i=(-1)^{t_i}$, t_i being the sum of the numbers of vertical steps in the h applications.

As an illustration of this theorem, we shall evaluate $\chi_{[4,2^2]}^{[4,2^2]}$. There are four ways in which the shape $[4, 2^2]$ can be built up by regular applications of 4, 2, 2 spaces in turn. These can be represented diagrammatically as follows:

1	1	3	3
1	2		
1	2		

,

1	1	2	2
1	3		
1	3		

,

1	1	1	1
2	3		
2	3		

,

1	1	1	1
2	2		
3	3		

.

The numbers in the spaces indicate the order in which the regular applications are made and do not of course refer to the letters upon which the permutations act.

The respective values of d_i are

$$(-1)^{2+1+0},\quad (-1)^{2+0+1},\quad (-1)^{0+1+1},\quad (-1)^{0+0+0}.$$

* Littlewood, p. 70.

It follows that

$$\chi^{[4,\,2^2]}_{[4,\,2^2]} = -1-1+1+1=0.$$

So long as the order in which the regular applications are made is the same in each process of building up the shape α, one may take the cycles β_i in any order. Thus in the last example there is no shape α' which yields $[4,\ 2^2]$ when a regular application of 4 spaces is applied to it. It follows immediately that $\chi^{[4,\,2^2]}_{[4,\,2^2]}=0$. One must be careful, however, not to count the first two diagrams above as one.

CHAPTER VI

SUBSTITUTIONAL EQUATIONS

Argument. In this chapter we return to the initial problem proposed by Young, that of solving equations of the type

$$LX = 0,$$

where L is a given substitutional expression and X is a substitutional expression to be determined. Young has shown that each such equation can be reduced to a set of matrix equations. There is in fact one such matrix equation for each partition α of n. In theory, at least, this solves the problem, but in practice the manipulation involved is enormous in all but the simplest cases. The theory can, however, be extended in the following way. Associated with any set of substitutional equations

$$L_1X = 0, \ldots, \quad L_rX = 0$$

we can construct a master idempotent J such that the solutions of the above set of equations are identical with those of

$$JX = 0.$$

Further, since J is an idempotent expression, the most general solution of the last equation is

$$X = (\epsilon - J)Y,$$

where Y is an arbitrary expression. The number of linearly independent solutions of this type is

$$n!(1 - j_1),$$

where j_1 is the coefficient of ϵ in J.

The foregoing methods may be extended to solve the set of equations

$$L_1X = R_1, \ldots, \quad L_rX = R_r,$$

where the expressions $L_1, \ldots, L_r, R_1, \ldots, R_r$ are given substitutional expressions.

§ 39. Minimum Functions

Let L be any given substitutional expression. The expressions $\epsilon, L, L^2, \ldots$ cannot all be linearly independent since they are each linear functions of the $n!$ permutations of $\mathcal{S}_n$. Let

$$\phi_L(x) = 0$$

be the equation of least degree which is satisfied by L. The polynomial $\phi_L(x)$ so defined is called the *minimum function* of L. Now if U_L^α be the matrix which represents L in the semi-normal representation α, the equation $\phi_L(x)=0$ must also be satisfied by U_L^α. That is, for every representation α

$$\phi_L(U_L^\alpha)=0.$$

The matrix U_L^α has, however, its own minimum function which we may denote by $\phi_L^\alpha(x)$ and which is usually called the reduced characteristic function of U_L^α. Thus, for every α

$$\phi_L^\alpha(U_L^\alpha)=0.$$

It follows that $\phi_L(x)$ contains each $\phi_L^\alpha(x)$ as a factor and so $\phi_L(x)$ contains the L.C.M. of the reduced characteristic functions $\phi_L^\alpha(x)$ as a factor. On the other hand, if $\Phi_L(x)$ be this L.C.M., the equation $\Phi_L(x)=0$ is satisfied by each matrix U_L^α and therefore by the substitutional expression L also. Thus, since $\phi_L(x)$ is the minimum function of L, $\Phi_L(x)$ must have the factor $\phi_L(x)$. It follows that, apart from a numerical factor, $\phi_L(x)$ and $\Phi_L(x)$ must be identical.

If for each α the matrix U_L^α is non-singular, then each matrix U_L^α possesses an inverse $(U_L^\alpha)^{-1}$, and so the expression L possesses an inverse L^{-1} such that

$$LL^{-1}=L^{-1}L=\epsilon.$$

This inverse L^{-1} is in fact the expression which is represented by the matrices $(U_L^\alpha)^{-1}$. Borrowing the nomenclature of matrices we shall term an expression L *non-singular* or *singular* according as it does or does not possess an inverse L^{-1}.

If L is a singular expression one or more of the matrices U_L^α must be singular, and the corresponding reduced characteristic functions $\phi_L^\alpha(x)$ must have a factor x possibly repeated.* Since $\phi_L(x)$ is the L.C.M. of these reduced characteristic functions, it too must possess a factor x possibly repeated. We may therefore write, when L is singular,

$$\phi_L(x)=x^i\psi_L(x),$$

where

$$\psi_L(0)\neq 0.$$

There is no lack of generality in stipulating that $\psi_L(0)=+1$ because $\phi_L(x)$ is so far undetermined to the extent of a non-zero numerical factor.

We shall now show that for any singular L it is possible to construct a non-singular expression A such that $\phi_{AL}(x)$ has the factor x unrepeated. Since L is singular at least some of the matrices U_L^α are singular. Let U_L be one of these. Since U_L possesses zero latent roots we may write †

$$U_L=H[P\dotplus Q_1\dotplus\ldots\dotplus Q_s]H^{-1},$$

where the middle factor is the canonical form of U_L, where P is the non-

* MacDuffee, p. 21.

† T. A., p. 61, or MacDuffee, p. 73.

singular submatrix corresponding to all the non-zero latent roots of U_L, and where each Q_i is a submatrix of order q_i of the form

$$Q_i = \begin{bmatrix} 0 & 1 & 0 & \dots & 0 \\ 0 & 0 & 1 & \dots & 0 \\ \dots & & & & \\ 0 & 0 & 0 & \dots & 1 \\ 0 & 0 & 0 & \dots & 0 \end{bmatrix}.$$

If we now write

$$U_A \equiv H[I \dotplus R_1 \dotplus \dots \dotplus R_s]H^{-1},$$

where I is a unit matrix of the same order as P and where R_i is a submatrix also of order q_i of the form

$$R_i = \begin{bmatrix} 0 & 0 & \dots & 0 & 1 \\ 1 & 0 & \dots & 0 & 0 \\ 0 & 1 & \dots & 0 & 0 \\ \dots & & & & \\ 0 & 0 & \dots & 1 & 0 \end{bmatrix},$$

then, by actual multiplication,

$$U_{AL} = U_A U_L = H[P \dotplus S_1 \dotplus \dots \dotplus S_s]H^{-1},$$

where S_i is a submatrix of order q_i of the form

$$S_i = \begin{bmatrix} 0 & 0 & 0 & \dots & 0 \\ 0 & 1 & 0 & \dots & 0 \\ 0 & 0 & 1 & \dots & 0 \\ \dots & & & & \\ 0 & 0 & 0 & \dots & 1 \end{bmatrix}.$$

Thus $P \dotplus S_1 \dotplus \dots \dotplus S_s$ is the canonical form of the matrix U_{AL} and each canonical submatrix associated with a zero latent root is of order unity. This means that any elementary divisor of the matrix U_{AL} corresponding to a zero latent root is linear and hence that the reduced characteristic function of U_{AL} must have the factor x unrepeated.

We can construct in the above manner a matrix U_A^α corresponding to each singular U_L^α, and each such matrix U_A^α is non-singular as is evident from the manner of its construction. On the other hand, if any U_L^α be non-singular, we define U_A^α to be the unit matrix. Hence if A be the substitutional expression represented by the matrices U_A^α, then A is non-singular, and $\phi_{AL}(x)$ being the

L.C.M. of the reduced characteristic functions of the matrices U^{α}_{AL} contains the factor x unrepeated. We are thus led to the following conclusion.*

THEOREM 24. *For any singular substitutional expression L a non-singular expression A can be constructed such that*

$$\phi_{AL}(x) = x\psi_{AL}(x), \quad \psi_{AL}(0) = 1.$$

Suppose now that M is any substitutional expression for which

$$\phi_M(x) = x\psi_M(x), \quad \psi_M(0) = 1.$$

Since $\psi_M(0) = 1$ it is clear that $1 - \psi_M(x)$ has a factor x and we may therefore write

$$1 - \psi_M(x) = x\theta_M(x).$$

Hence

$$\{\epsilon - \psi_M(M)\}\psi_M(M) = \theta_M(M)M\psi_M(M) = \theta_M(M)\phi_M(M) = 0,$$

from which it follows that

$$\{\psi_M(M)\}^2 = \psi_M(M).$$

Consequently,

$$\{\epsilon - \psi_M(M)\}^2 = \epsilon - \psi_M(M).$$

These results may be expressed as follows.

THEOREM 25. *If the minimum function of M is $x\psi_M(x)$, where $\psi_M(0) = 1$, then the expressions*

$$\psi_M(M), \quad \epsilon - \psi_M(M)$$

are idempotents.

Theorems 24 and 25 taken together demonstrate how an idempotent expression $\psi_{AL}(AL)$ can be constructed from any singular expression L. This idempotent we shall refer to as an idempotent *associated* with L. It should be pointed out, however, that the non-singular expression A referred to in the enunciation of theorem 24 is by no means unique. In the proof of this theorem a method is given whereby an expression A can be constructed when no suitable A can be found by inspection. In most simple cases, however, an A can be found quite easily by inspection or even by trial and error. When any such A has been found, the expression $\psi_{AL}(AL)$ gives an idempotent associated with L.

§ 40. The Master Idempotent

Suppose now that $L_1, L_2, \ldots, L_r$ are a given set of singular substitutional expressions. To avoid unnecessary complexity in the formulae which we are about to use we shall modify our notation slightly and denote the idempotents associated with certain expressions $L_1, L_{(2)}, \ldots, L_{(r)}$ by $\psi_1, \psi_{(2)}, \ldots, \psi_{(r)}$ respectively. We also write

$$\psi_1\psi_{(2)} \ldots \psi_{(i)} \equiv K_i.$$

* Rutherford II, p. 121.

The expressions $L_{(2)}, \ldots, L_{(r)}$ have still to be defined and we do this inductively by the formulae

$$L_{(2)} \equiv L_2 K_1, \quad L_{(3)} \equiv L_3 K_2, \ldots, \quad L_{(r)} \equiv L_r K_{r-1}.$$

We now prove by induction that

$$L_j K_i = 0. \quad (j = 1, \ldots, i). \qquad \ldots (40.1)$$

The basis of the induction is

$$L_1 K_1 = L_1 \psi_1 = \phi_1 = 0.$$

Since $K_i = K_{i-1}\psi_{(i)}$, the induction hypothesis gives

$$L_j K_i = L_j K_{i-1}\psi_{(i)} = 0,$$

when $j < i$. On the other hand,

$$L_i K_i = L_i K_{i-1}\psi_{(i)} = L_{(i)}\psi_{(i)} = \phi_{(i)} = 0.$$

The formula (40.1) has therefore been established.

We can also prove by induction that K_i is an idempotent, the basis for this being the fact that

$$K_1^2 = \psi_1^2 = \psi_1 = K_1.$$

We assume, therefore, that $K_{i-1}^2 = K_{i-1}$. Now $\psi_{(i)}$ is known to be of the form

$$\psi_{(i)} = \epsilon - \theta_{(i)}(A_{(i)}L_{(i)})A_{(i)}L_{(i)},$$

and $L_{(i)} = L_i K_{i-1}$. We therefore deduce that

$$\begin{aligned}\psi_{(i)}K_{i-1} &= K_{i-1} - \theta_{(i)}(A_{(i)}L_{(i)})A_{(i)}L_i K_{i-1}^2 \\ &= K_{i-1} - \theta_{(i)}(A_{(i)}L_{(i)})A_{(i)}L_i K_{i-1} \\ &= K_{i-1} + \psi_{(i)} - \epsilon.\end{aligned}$$

Hence, since K_{i-1} and $\psi_{(i)}$ are idempotents,

$$K_i^2 = K_{i-1}\psi_{(i)}K_{i-1}\psi_{(i)} = K_{i-1}\{K_{i-1} + \psi_{(i)} - \epsilon\}\psi_{(i)} = K_{i-1}\psi_{(i)} = K_i.$$

Since each K_i is an idempotent, so also is each expression $\epsilon - K_i$. We shall call $\epsilon - K_r$ the *master idempotent* associated with the set of singular expressions $L_1, \ldots, L_r$ and we shall denote it by J. This master idempotent has the following properties:

$$J^2 = J,$$

$$L_i J = L_i,$$

the second of which follows immediately from (40.1).

§ 41. The Substitutional Properties of Functions

In general, when we operate with a permutation σ of $\mathcal{S}_n$, on a one-valued function $F(z_1, \ldots, z_n)$ of the parameters $z_1, \ldots, z_n$, we obtain a new function F_σ of the same n parameters. We express this fact by writing

$$\sigma F = F_\sigma.$$

For certain functions F, however, and for certain permutations σ it may happen that

$$F_\sigma = kF,$$

where k is numerical. In such a case

$$(\sigma - k\epsilon)F = 0.$$

More generally, there may be substitutional expressions

$$L \equiv \Sigma l_i \sigma_i,$$

where the coefficients l_i are numerical, such that

$$LF = 0.$$

This equation expresses the fact that the function F has certain substitutional properties. If no such equation holds then F has no substitutional properties.

Consider now the $n!$ functions F_{σ_i} obtained by operating on F with the $n!$ permutations σ_i. If these $n!$ functions are linearly independent, F cannot satisfy any relation $LF = 0$ except in the trivial case where L is identically zero. If, however, the functions F_{σ_i} are not linearly independent, their dependence can be expressed by one or more relations of the type

$$l_1 F + l_2 F_{\sigma_2} + \ldots + l_{n!} F_{\sigma_{n!}} = 0,$$

where the l_i are numerical. Since $F_{\sigma_i} = \sigma_i F$, the foregoing relation may be written

$$LF = 0,$$

where, as before, $L \equiv \Sigma l_i \sigma_i$. Thus all the substitutional properties of any given F may be expressed in terms of a set of equations of the form

$$L_1 F = 0, \quad L_2 F = 0, \ldots, \quad L_r F = 0.$$

Suppose now that F is an unknown function which satisfies these relations. If any L_i is non-singular, then

$$F = L_i^{-1} L_i F = 0,$$

and $F = 0$ is the only solution of the above set of relations. We may therefore confine our attention to the case where each L_i is singular and we shall now prove the following theorem.

THEOREM 26. *If $L_1, \ldots, L_r$ is a set of singular substitutional expressions then the set of equations*

$$L_1 F = 0, \ldots, \quad L_r F = 0 \qquad \ldots(41.1)$$

is equivalent to the single equation

$$JF = 0, \qquad \ldots(41.2)$$

where J is the master idempotent of the set.

PROOF. The proof consists of two parts. First, we shall show that the equations (41.1) are deducible from equation (41.2) and, secondly, we shall show that equation (41.2) is deducible from the set of equations (41.1).

(i) Since we have shown in the last section that

$$L_iJ = L_i,$$

it is clear that if $JF = 0$, then

$$L_iF = L_iJF = 0$$

for all values of i.

(ii) On the other hand, if, using the notation of § 40, we assume the validity of the equations

$$(\epsilon - K_{i-1})F = 0, \quad L_iF = 0,$$

then

$$L_{(i)}F = (L_{(i)} - L_i)F = L_i(K_{i-1} - \epsilon)F = 0.$$

Now $\epsilon - \psi_{(i)}$ has a factor $A_{(i)}L_{(i)}$ and therefore, since $L_{(i)}F = 0$,

$$(\epsilon - \psi_{(i)})F = 0.$$

Thus

$$F = \psi_{(i)}F = \psi_{(i)}K_{i-1}F = K_iF.$$

An induction has therefore been established which shows that the formula

$$(\epsilon - K_i)F = 0$$

is a consequence of the formulae

$$(\epsilon - K_{i-1})F = 0, \quad L_iF = 0.$$

Now, since $L_1F = 0$, then

$$(\epsilon - K_1)F = (\epsilon - \psi_1)F = \theta_1A_1L_1F = 0.$$

This provides a basis of the induction which proves that

$$(\epsilon - K_i)F = 0$$

is a consequence of the i equations

$$L_1F = 0, \ldots, \quad L_iF = 0.$$

In particular, since $J = \epsilon - K_r$, the relation

$$JF = 0$$

is a consequence of the r equations

$$L_1F = 0, \ldots, \quad L_rF = 0.$$

The importance of this theorem lies in the fact that we can replace any set (41.1) of substitutional equations where the substitutional operators are all singular by a single equation of which the operator is an idempotent.

Since $JF=0$, we have

$$F=(\epsilon-J)F,$$

and since $J=J^2$, the equation $JF=0$ may be regarded as a consequence of $J(\epsilon-J)=0$. That is, by expressing F in the form $(\epsilon-J)F$ we can transfer all the substitutional properties of the function F to the substitutional operator $\epsilon-J$. We therefore have the following result.*

THEOREM 27. *If F is any one valued function of the parameters $z_1, \ldots, z_n$ we can find a substitutional expression X (namely, the expression $\epsilon-J$) such that*

$$F=XF,$$

and such that any substitutional equation $LX=0$ is satisfied if, and only if, the functional relation $LF=0$ is satisfied.

In virtue of this theorem we can replace any equation of the form $LF=0$, where the L is a known expression and F is to be determined, by a substitutional equation $LX=0$, where X is an unknown substitutional expression. When the most general solution X of $LX=0$ has been determined, the most general solution of $LF=0$ is then

$$F(z_1, \ldots, z_n)=XC(z_1, \ldots, z_n),$$

where C is an arbitrary one-valued function of the parameters $z_1, \ldots, z_n$. The remainder of this chapter will therefore be concerned mainly with substitutional equations such as $LX=0$ rather than with functional equations such as $LF=0$.

§ 42. The Number of Independent Solutions of $LX=0$

Our first problem is that of solving the equation

$$LX=0, \qquad \ldots(42.1)$$

where L is a known and X is an unknown substitutional expression. If we write

$$L=\sum_i l_{\sigma_i}\sigma_i, \quad X=\sum_i x_{\sigma_i}\sigma_i,$$

where l_{σ_i} and x_{σ_i} are numerical, then

$$LX=\sum_{i,j} l_{\sigma_i}x_{\sigma_j}\sigma_i\sigma_j=\sum_{k,j} l_{\sigma_k\sigma_j^{-1}}x_{\sigma_j}\sigma_k.$$

Since the permutations σ_k are linearly independent, we conclude that the equations †

$$\sum_j l_{\sigma_k\sigma_j^{-1}}x_{\sigma_j}=0, \quad (k=1, \ldots, n!) \qquad \ldots(42.2)$$

must hold if $LX=0$. Evidently the number of linearly independent solutions of these equations depends upon the rank of the matrix Λ which has $l_{\sigma_k\sigma_j^{-1}}$ for its k,jth element. If Λ is of rank $n!$, the only solution of the equations

* Young I, p. 102. † Young I, p. 104.

(42.2) is $x_{\sigma_j}=0$ for all j, and the only solution of $LX=0$ is $X=0$. If, however, the rank λ of Λ is less than $n!$, there are $n!-\lambda$ linearly independent non-zero solutions of (42.2), and the most general solution of (42.1) is a linear combination of $n!-\lambda$ independent solutions.

A more fruitful method of attack is to express L and X in terms of the semi-normal units e_{rs}^{α}. The natural or orthogonal units would do equally well for our purpose but we shall use the semi-normal ones here. Let

$$L=\sum_{\alpha,r,s} u_{rsL}^{\alpha}e_{rs}^{\alpha},\quad X=\sum_{\alpha,r,s} u_{rsX}^{\alpha}e_{rs}^{\alpha}.$$

As a consequence of $LX=0$, we have

$$U_L^{\alpha}U_X^{\alpha}=0$$

for every representation α. It will be seen that the equations for the elements of any column of U_X^{α} are the same as those for any other column. Thus, if u_X^{α} denote a typical column, the equations for its elements are comprised in the single matrix equation

$$U_L^{\alpha}u_X^{\alpha}=0.$$

If the rank of U_L^{α} is λ^{α}, then there are $f^{\alpha}-\lambda^{\alpha}$ independent parameters in the most general solution for u_X^{α}. In fact, any solution u_X^{α} may be expressed in the form

$$u_X^{\alpha}=\sum_i \xi_i^{\alpha}u_{X_i}^{\alpha},$$

in which $u_{X_i}^{\alpha}$ are the $f^{\alpha}-\lambda^{\alpha}$ independent solutions and the ξ_i^{α} are a like number of arbitrary numerical coefficients. The coefficients ξ_i^{α} will in general take different values for the different column vectors u_X^{α}, but the $u_{X_i}^{\alpha}$ may be taken to be the same for all columns of U_X^{α}. There are therefore * in all $f^{\alpha}(f^{\alpha}-\lambda^{\alpha})$ arbitrary parameters in the most general solution of $U_L^{\alpha}U_X^{\alpha}=0$. It follows that the number of linearly independent solutions of the equation $LX=0$ is

$$\sum_{\alpha} f^{\alpha}(f^{\alpha}-\lambda^{\alpha}),$$

or, since $\sum_{\alpha}(f^{\alpha})^2=n!$, the number is

$$n!-\sum_{\alpha} f^{\alpha}\lambda^{\alpha}. \qquad \ldots(42.3)$$

In particular, if L is idempotent, that is, if $L^2=L$, then the matrix U_L^{α} is also idempotent. Now it is known that the rank of an idempotent matrix is equal to its trace ; hence, if $L=\sum_i l_i\sigma_i$, the rank of U_L^{α} is

$$\lambda^{\alpha}=\text{tr. } U_L^{\alpha}=\sum_i l_i(\text{tr. } U_{\sigma_i}^{\alpha})=\sum_i l_i\chi_{\sigma_i}^{\alpha},$$

for the trace of $U_{\sigma_i}^{\alpha}$ is simply the character component $\chi_{\sigma_i}^{\alpha}$. Further, by (35.4), $\sum_{\alpha} f^{\alpha}\chi_{\sigma_i}^{\alpha}$ is equal to $n!$ or 0 according as σ_i is or is not the unit permutation ϵ, and so

$$\sum_{\alpha} f^{\alpha}\lambda^{\alpha}=\sum_{\alpha,i} f^{\alpha}l_i\chi_{\sigma_i}^{\alpha}=n!\,l_1,$$

* Young III, p. 267.

where l_1 is the coefficient of ϵ in L. Substituting this value in (42.3) we obtain the following result.

THEOREM 28. *If L be an idempotent substitutional expression, the number of linearly independent solutions of* $LX=0$ *is* $n!(1-l_1)$, *where* l_1 *is the coefficient of* ϵ *in L.*

Now, if L is idempotent, so is U_L^α. It follows from this that

$$U_X^\alpha = U_\epsilon^\alpha - U_L^\alpha = U_{\epsilon-L}^\alpha$$

is one solution of the equation

$$U_L^\alpha U_X^\alpha = 0.$$

Again, if λ^α be the rank of U_L^α, the rank of $U_{\epsilon-L}^\alpha$ is just $f^\alpha - \lambda^\alpha$. There are therefore $f^\alpha - \lambda^\alpha$ linearly independent columns in the matrix $U_{\epsilon-L}^\alpha$, and this is exactly the number of linearly independent solutions of the equation

$$U_L^\alpha u_X^\alpha = 0.$$

Each column of $U_{\epsilon-L}^\alpha$ is a solution of this last equation and any linear combination of these solutions may be written in the form $U_{\epsilon-L}^\alpha u_Y^\alpha$, where u_Y^α is a suitably chosen column vector. The most general solution of the last equation is therefore expressible in the form

$$u_X^\alpha = U_{\epsilon-L}^\alpha u_Y^\alpha,$$

where u_Y^α is now an arbitrary column vector. Similarly, the most general solution of $U_L^\alpha U_X^\alpha = 0$ is

$$U_X^\alpha = U_{\epsilon-L}^\alpha U_Y^\alpha,$$

where U_Y^α is an arbitrary matrix. Since this is true for every representation α, we are led to the following result.*

THEOREM 29. *If L be an idempotent substitutional expression, the most general solution of the equation* $LX=0$ *is*

$$X = (\epsilon - L)Y,$$

where Y is an arbitrary substitutional expression.

§ 43. The Solution of the Equation $LX=0$

By means of this last theorem we are now able to find the most general solution of the substitutional equation

$$LX = 0. \qquad \ldots(43.1)$$

If L is non-singular, $X=0$ is the only solution, for in this case L^{-1} exists aud

$$X = L^{-1}LX = 0.$$

* Rutherford II, p. 119.

We may therefore suppose that L is singular. This being so, we can construct as in theorem 24 (§ 39) a non-singular expression A such that

$$\phi_{AL}(x)=x\psi_{AL}(x), \quad \psi_{AL}(0)=1.$$

Theorem 25 (§ 39) then informs us that the expressions

$$\psi_{AL}(AL), \quad \epsilon-\psi_{AL}(AL)$$

are idempotents. Also, since $\psi_{AL}(0)=1$, the idempotent $\epsilon-\psi_{AL}(AL)$ has a factor AL. In view of this fact it should be clear that any solution of (43.1) is also a solution of

$$\{\epsilon-\psi_{AL}(AL)\}X=0. \qquad \ldots(43.2)$$

On the other hand, since

$$L\psi_{AL}(AL)=A^{-1}AL\psi_{AL}(AL)=A^{-1}\phi(AL)=0,$$

we have

$$L\{\epsilon-\psi_{AL}(AL)\}=L,$$

from which we deduce that every solution of (43.2) is also a solution of (43.1). We conclude, therefore, that the equations (43.1) and (43.2) have exactly the same solutions. Since, however, $\epsilon-\psi_{AL}(AL)$ is an idempotent, we deduce from theorem 29 (§ 42) that the most general solution of (43.2) and therefore of (43.1) is

$$X=\psi_{AL}(AL)Y, \qquad \ldots(43.3)$$

where Y is an arbitrary expression.

The solution of a set of simultaneous substitutional equations of the form

$$L_1X=0, \ldots, \quad L_rX=0$$

is merely a generalisation of the preceding argument in which the idempotent $\epsilon-\psi_{AL}(AL)$ is replaced by the master idempotent J of the set $L_1, \ldots, L_r$.

If any of the given expressions L_i is non-singular, then we have

$$X=L_i^{-1}L_iX=0,$$

and this is the only solution.

If every L_i is singular, then just as in theorem 26 (§ 41) we can replace the set of equations $L_iX=0$ by the single equation

$$JX=0,$$

where J is the master idempotent of the set $L_1, \ldots, L_r$. By theorem 29 (§ 42) the most general solution of this equation is

$$X=(\epsilon-J)Y,$$

where Y is an arbitrary expression.

In the above solution there are $n!$ arbitrary parameters, namely the coefficients in Y of the $n!$ permutations of $\mathcal{S}_n$. Nevertheless, as we have seen in theorem 28 (§ 42), there are only $n!(1-j_1)$ independent solutions, where j_1 is the coefficient of ϵ in J. In the general case it may be difficult to pick out $n!(1-j_1)$ solutions which are linearly independent from the $n!$ solutions

$$X=(\epsilon-J)\sigma_i, \quad (i=1,\ldots,n!)$$

which we have just obtained. It is, however, possible to do this in certain special cases which are of some importance. This will be done in § 45.

We shall conclude this section by illustrating the preceding theory by means of a particular example when $n=4$.

Let

$$L_1=\epsilon-(\mathbf{1},\mathbf{2}), \quad L_2=\epsilon-(\mathbf{3},\mathbf{4}).$$

Since $L_1^2=2L_1$, we find that

$$\psi_1=-\tfrac{1}{2}L_1+\epsilon=\tfrac{1}{2}(\epsilon+(\mathbf{1},\mathbf{2})).$$

The most general solution of the equation $L_1X=0$ is therefore

$$X=\tfrac{1}{2}(\epsilon+(\mathbf{1},\mathbf{2}))Y,$$

where Y is arbitrary. Now the coefficient of ϵ in $\epsilon-\psi_1$, that is, in

$$\epsilon-\tfrac{1}{2}(\epsilon+(\mathbf{1},\mathbf{2})),$$

is $\frac{1}{2}$. The number of linearly independent solutions is therefore $4!(1-\frac{1}{2})$, i.e. 12. Now the $4!$ permutations of $\mathcal{S}_4$ can be separated into associate complexes with respect to the sub-group whose elements are ϵ, $(\mathbf{1},\mathbf{2})$. That is, $4!/2$ permutations $\epsilon,\ldots,\tau_{4!/2}$ can be found such that

$$\epsilon,\ldots,\tau_{4!/2}, (\mathbf{1},\mathbf{2}),\ldots,(\mathbf{1},\mathbf{2})\tau_{4!/2}$$

are the $4!$ elements of $\mathcal{S}_4$. Also,

$$\tfrac{1}{2}(\epsilon+(\mathbf{1},\mathbf{2}))(\mathbf{1},\mathbf{2})\tau_i=\tfrac{1}{2}(\epsilon+(\mathbf{1},\mathbf{2}))\tau_i.$$

It follows that of the $4!$ solutions

$$X=\tfrac{1}{2}(\epsilon+(\mathbf{1},\mathbf{2}))\sigma_k$$

only the 12 solutions

$$X=\tfrac{1}{2}(\epsilon+(\mathbf{1},\mathbf{2}))\tau_i$$

are independent. The 12 solutions just obtained are not all solutions of $L_2X=0$. To find the most general solution of the two equations $L_1X=0$, $L_2X=0$, we must first construct the master idempotent J. We write

$$L_{(2)}\equiv L_2\psi_1=\tfrac{1}{2}(\epsilon+(\mathbf{1},\mathbf{2})-(\mathbf{3},\mathbf{4})-(\mathbf{1},\mathbf{2})(\mathbf{3},\mathbf{4})).$$

It is then easily verified that $L_{(2)}^2=2L_{(2)}$ and that therefore

$$\psi_{(2)}=-\tfrac{1}{2}L_{(2)}+\epsilon=\epsilon-\tfrac{1}{4}(\epsilon+(\mathbf{1},\mathbf{2})-(\mathbf{3},\mathbf{4})-(\mathbf{1},\mathbf{2})(\mathbf{3},\mathbf{4})).$$

Thus,

$$\epsilon-J\equiv\psi_1\psi_{(2)}=\tfrac{1}{4}(\epsilon+(\mathbf{1},\mathbf{2})+(\mathbf{3},\mathbf{4})+(\mathbf{1},\mathbf{2})(\mathbf{3},\mathbf{4})).$$

The most general solution of $JX=0$ is therefore

$$X=\tfrac{1}{4}(\epsilon+(1,2)+(3,4)+(1,2)(3,4))Y,$$

where Y is arbitrary. Of the 4! solutions

$$X=\tfrac{1}{4}(\epsilon+(1,2)+(3,4)+(1,2)(3,4))\sigma_k$$

implied in this last equation only $4!/4$ of them are linearly independent. Since ϵ, (1, 2), (3, 4), (1, 2)(3, 4) are the elements of a sub-group of $\mathcal{S}_4$ we can resolve $\mathcal{S}_4$ into associate complexes as before and find $4!/4$ permutations $\epsilon, \ldots, \tau_{4!/4}$ such that the 6 independent solutions of $L_1X=0$, $L_2X=0$ may be taken to be

$$X=\tfrac{1}{4}(\epsilon+(1,2)+(3,4)+(1,2)(3,4))\tau_i.$$

We could of course dispense with the numerical factor $\tfrac{1}{4}$ at this stage but its presence is necessary if we wish to emphasise the idempotent properties of the expression

$$\tfrac{1}{4}(\epsilon+(1,2)+(3,4)+(1,2)(3,4)).$$

§ 44. Functions which are Invariant under a Group of Substitutional Expressions

Before proceeding to discuss other substitutional equations we shall mention one or two important applications of the results we have already achieved.

If the substitutional expressions $L_1, L_2, \ldots, L_g$ are the elements of a group of order g and if we write

$$G\equiv(1/g)(L_1+\ldots+L_g),$$

then from the well-known group property that

$$L_iG=G$$

it is clear that $G^2=G$. That is to say, G, and therefore $\epsilon-G$, are idempotents. The most general solution of $GX=0$ is therefore $X=(\epsilon-G)Y$ and that of $(\epsilon-G)X=0$ is $X=GY$, where in both cases Y is arbitrary. Now if $(\epsilon-G)X=0$, we have

$$X=GX$$

and therefore

$$L_iX=L_iGX=GX=X.$$

Any solution X is therefore invariant under the application of each element of the group. Conversely, if for every L_i we have

$$L_iX=X,$$

then

$$GX=(1/g)\Sigma L_iX=(1/g)gX=X.$$

The necessary and sufficient condition that X be invariant under the application of each element L_i of a group is therefore that

$$(\epsilon-G)X=0,$$

and the most general expression X with this property is GY where Y is arbitrary.

Suppose now that the substitutional properties of a function F are specified by the fact that for each permutation of the set $\epsilon, \tau_2, \ldots, \tau_g$ and for no other permutation we have

$$\tau_i F = \omega_i F,$$

where ω_i is numerical and non-zero. In such a case the permutations ϵ, $\tau_2, \ldots, \tau_g$ form a sub-group $\mathcal{G}$ of $\mathcal{S}_n$, as will now be shown. (i) If $\tau_i F = \omega_i F$ and $\tau_j F = \omega_j F$, then $\tau_i \tau_j F = \omega_j \tau_i F = \omega_i \omega_j F$, from which we see that the product of two elements of the set is an element of the set. (ii) Since the elements are permutations the associative law holds. (iii) Since $\epsilon F = F$, the set contains the unit element ϵ. (iv) If $\tau_i F = \omega_i F$, then $F = \omega_i^{-1} \tau_i F$ and so $\tau_i^{-1} F = \tau_i^{-1} \omega_i^{-1} \tau_i F = \omega_i^{-1} F$. It follows that the inverse of every element of the set is an element of the set. Since we have just shown that the elements satisfy the four group postulates, we conclude that the set forms a group $\mathcal{G}$ which is a sub-group of $\mathcal{S}_n$.

It is easy to see that the numerical factors $\omega_1, \omega_2, \ldots, \omega_g$ have the same multiplication table as the permutations $\epsilon, \tau_2, \ldots, \tau_g$. This means that

$$\tau_i : \omega_i$$

is a representation of the sub-group $\mathcal{G}$. This may be regarded as a matrix representation of order unity. Since it is well known in group theory that the gth power of any element of a group of order g is the unit element, we come to the conclusion that

$$\omega_i^g = 1.$$

Each ω is therefore a gth root of unity.

If we now write

$$L_i \equiv \omega_i^{-1} \tau_i,$$

it will be seen that the expressions L_i are the elements of a group which is simply isomorphic with the group $\mathcal{G}$ of permutations. In fact

$$\tau_i : L_i$$

is another representation of $\mathcal{G}$ although it is not this time a matrix representation. Since the substitutional properties of F are specified by the equations

$$L_i F = F,$$

we have only to solve the substitutional equations

$$L_i X = X$$

and we know that the most general solution of these equations is

$$X = GY,$$

where Y is arbitrary and where

$$G \equiv (1/g)(L_1 + \ldots + L_g) = (1/g)(\epsilon + \omega_2^{-1}\tau_2 + \ldots + \omega_g^{-1}\tau_g).$$

The most general function F with the required properties is therefore

$$F = GC,$$

where C is an arbitrary function of the n parameters $z_1, \ldots, z_n$. The factor $(1/g)$ may of course be incorporated in the arbitrary Y or C if desired.

§ 45. Some Special Cases

We have seen in the previous section that if the permutations $\epsilon, \tau_2, \ldots, \tau_g$ form a sub-group $\mathcal{G}$ of $\mathcal{S}_n$, then the most general solution of the equations

$$\tau_i X = \omega_i X$$

may be expressed in the form

$$X = (1/g)(\epsilon + \omega_2^{-1}\tau_2 + \ldots + \omega_g^{-1}\tau_g)Y, \qquad \ldots(45.1)$$

where Y is arbitrary. This is in fact the most general solution of

$$(\epsilon - G)X = 0,$$

where

$$G = (1/g)(\epsilon + \omega_2^{-1}\tau_2 + \ldots + \omega_g^{-1}\tau_g).$$

Since $\epsilon - G$ is an idempotent, we know by theorem 28 (§ 42) that the number of linearly independent solutions is $n!\,(1 - l_1)$ where l_1 is the coefficient $1 - (1/g)$ of ϵ in $\epsilon - G$. There are therefore in all exactly $n!/g$ linearly independent solutions.

Since $\mathcal{G}$ is a sub-group of $\mathcal{S}_n$, we can find $n!/g$ permutations $\epsilon, \rho_2, \ldots, \rho_{n!/g}$ such that the $n!$ permutations

$$\tau_i \rho_j \quad (i = 1, \ldots g\,;\ \ j = 1, \ldots, n!/g)$$

are all different and therefore comprise all the permutations of $\mathcal{S}_n$. Of the permutations ρ_i only ϵ is an element of $\mathcal{G}$. If we now give Y in (45.1) the values $\epsilon, \rho_2, \ldots, \rho_{n!/g}$ in turn we obtain just $n!/g$ solutions and these must be linearly independent, since any dependence relation would imply a relation between the independent permutations of $\mathcal{S}_n$. We may therefore state that the most general solution of the equations

$$\tau_i X = \omega_i X$$

or of the single equation

$$(\epsilon - G)X = 0$$

is a linear combination of the independent solutions

$$X = G\rho_i, \quad (i = 1, \ldots, n!/g).$$

Again, the most general solution of

$$GX=0$$

is

$$X=(\epsilon-G)Y,$$

where Y is arbitrary. Further, since the coefficient of ϵ in G is $1/g$, there are in all $n!(1-(1/g))$ independent solutions. It is easily demonstrated that there are just this number of solutions of the form

$$(\epsilon-G)\tau_i\rho_j, \quad (i=2,\ldots,g;\ \ j=1,\ldots,n!/g)$$

and we shall show that these are independent. As we have seen, there can be no linear relation between solutions of this type which involve different ρ_j. It is therefore sufficient to show that no relation

$$\sum_{i=2}^{g} k_i(\epsilon-G)\tau_i=0$$

exists unless each numerical coefficient k_i vanishes. Since $G\tau_i=\omega_i GL_i=G\omega_i$, such a relation would give

$$\sum_{i=2}^{g} k_i\tau_i=\left(\sum_{i=2}^{g} k_i\omega_i\right)G.$$

Equating the coefficients of ϵ on both sides, we find that

$$\sum_{i=2}^{g} k_i\omega_i=0,$$

and so

$$\sum_{i=2}^{g} k_i\tau_i=0.$$

Since the permutations τ_i are linearly independent we conclude that each k_i vanishes.

Alternative independent solutions of $GX=0$ are

$$(\epsilon-\omega_i^{-1}\tau_i)\rho_j, \quad (i=2,\ldots,g;\ \ j=1,\ldots,n!/g).$$

They are linearly independent since the permutations of $\mathcal{S}_n$ are independent. Since $G\tau_i=G\omega_i$, each of these solutions satisfies the equation

$$GX=0.$$

Suppose next that L is a given substitutional expression and that the group generated by the permutations which appear in L with non-zero coefficients is of order g. If we denote the elements of this group by $\epsilon, \tau_2, \ldots, \tau_g$, then we may write

$$L=\Sigma\, l_{1i}\tau_i.$$

It is not of course necessary that each coefficient l_{1i} be non-zero.

In general, the minimum function $\phi_L(x)$ of L is of degree less than or equal to g but we here restrict our attention to the case where $\phi_L(x)$ is of degree g. We therefore have $g-1$ independent equations

$$L^j = (\Sigma l_{1i}\tau_i)^j = \Sigma l_{ji}\tau_i, \quad (j=1, \ldots, g-1)$$

which may be solved for the permutations $\tau_2, \ldots, \tau_g$. Each τ_i is therefore expressible as a polynomial in L of degree $g-1$ at most. We may therefore write

$$\tau_i = \lambda_{i0}\epsilon + \lambda_{i1}L + \ldots + \lambda_{i,g-1}L^{g-1}, \quad (i=2, \ldots, g)$$

where the λ_{ij} are numerical coefficients. In the case under consideration the equation

$$LX = 0$$

leads to

$$\tau_i X = \lambda_{i0}X, \quad (i=1, \ldots, g)$$

where we define $\lambda_{10} = +1$. This is the case considered earlier in this section and the most general solution is

$$X = (1/g)(\epsilon + \lambda_{20}^{-1}\tau_2 + \ldots + \lambda_{g0}^{-1}\tau_g)Y,$$

where Y is arbitrary, and where the factor $1/g$ can be omitted if desired.

In illustration we consider the case

$$\begin{aligned} L &= \epsilon - (1, 2, 3), \\ L^2 &= \epsilon - 2(1, 2, 3) + (3, 2, 1), \\ L^3 &= \quad -3(1, 2, 3) + 3(3, 2, 1), \\ L^3 &- 3L^2 + 3L = 0. \end{aligned}$$

Solving these equations, we obtain

$$\begin{aligned} \epsilon &= \epsilon, \\ (1, 2, 3) &= \epsilon - L, \\ (3, 2, 1) &= \epsilon - 2L + L^2, \end{aligned}$$

and so $\lambda_{i0} = +1$ for each case. The most general solution of $LX = 0$ is therefore

$$X = (\epsilon + (1, 2, 3) + (3, 2, 1))Y.$$

§ 46. The Equation $LX = R$

We now turn our attention to the substitutional equation

$$LX = R,$$

where L and R are known substitutional expressions and X is to be determined. If L is non-singular there is a unique solution,

$$X = L^{-1}R.$$

If L is singular we can construct as in theorem 24 (§ 39) a non-singular expression A such that

$$\phi_{AL}(x) = x\psi_{AL}(x) = x(1 - x\theta_{AL}(x)).$$

The equation

$$ALX = AR \qquad \ldots(46.1)$$

has evidently the same solutions as the equation $LX = R$. Since

$$\phi_{AL}(AL) = \psi_{AL}(AL)AL = 0,$$

it is clear that the equation $LX = R$ has no solutions whatever unless

$$\psi_{AL}(AL)AR = 0.$$

On the other hand, if R satisfies this last equation,

$$\{\epsilon - AL\theta_{AL}(AL)\}AR = 0\,;$$

that is,

$$AL\theta_{AL}(AL)AR = AR.$$

Comparing this equation with (46.1), we see that

$$X = \theta_{AL}(AL)AR \qquad \ldots(46.2)$$

is a particular solution. A criterion has therefore been established for the existence of solutions of the equation $LX = R$, where L is singular.*

THEOREM 30. *The necessary and sufficient condition that the equation $LX = R$ should have a solution, L being singular, is that*

$$\psi_{AL}(AL)AR = 0.$$

If X and X_1 be any two solutions of (46.1), the expression $X - X_1$ must satisfy †

$$AL(X - X_1) = 0,$$

and so, analogous to (43.3), we have

$$X - X_1 = \psi_{AL}(AL)Y$$

for some value of Y. If we now take X_1 to be the particular solution (46.2) already found, we see that any solution X may be expressed in the form

$$X = \theta_{AL}(AL)AR + \psi_{AL}(AL)Y.$$

This is in fact the most general solution of (46.1) if Y is regarded as an arbitrary expression.

It is also possible to solve a set of simultaneous equations

$$L_1X = R_1, \ldots, \quad L_rX = R_r. \qquad \ldots(46.3)$$

* Rutherford II, p. 124. † Young I, p. 106.

If any one of these equations has no solution there is no more to be said, while if any L_i is non-singular, there is a unique solution,

$$X = L_i^{-1} R_i,$$

provided this value is consistent with the other equations. If each L_i is singular we can first solve the equation $L_1 X = R_1$ and write the solution in the form

$$X = \theta_1 A_1 R_1 + \psi_1 Y,$$

where θ_1 and ψ_1 stand for $\theta_{A_1 L_1}(A_1 L_1)$ and $\psi_{A_1 L_1}(A_1 L_1)$. The expression Y is not necessarily arbitrary if X is subject to a second condition $L_2 X = R_2$ because it must satisfy the equation

$$L_2 \theta_1 A_1 R_1 + L_2 \psi_1 Y = R_2.$$

This last equation may, however, be written

$$L_{(2)} Y = R_{(2)},$$

where $L_{(2)} = L_2 \psi_1$ and $R_{(2)} = R_2 - L_2 \theta_1 A_1 R_1$. This equation in turn may have no solution, but if $L_{(2)}$ is singular and $\psi_{(2)} A_{(2)} R_{(2)} = 0$ we obtain

$$Y = \theta_{(2)} A_{(2)} R_{(2)} + \psi_{(2)} Z$$

and therefore

$$X = \theta_1 A_1 R_1 + \psi_1 (\theta_{(2)} A_{(2)} R_{(2)} + \psi_{(2)} Z).$$

A third equation $L_3 X = R_3$ implies further conditions which must be satisfied by Z, and so on. Eventually after r steps the most general solution of the equations (46.3) is achieved.

APPENDIX

TABLES FOR THE CASE $n=3$

THE elements of the group $\mathfrak{S}_3$ are

$$\epsilon,\ (2, 3),\ (3, 1),\ (1, 2),\ (1, 2, 3),\ (3, 2, 1).$$

The group multiplication table is :

	ϵ	(2, 3)	(3, 1)	(1, 2)	(1, 2, 3)	(3, 2, 1)
ϵ	ϵ	(2, 3)	(3, 1)	(1, 2)	(1, 2, 3)	(3, 2, 1)
$(2, 3)^{-1}=(2, 3)$	(2, 3)	ϵ	(1, 2, 3)	(3, 2, 1)	(3, 1)	(1, 2)
$(3, 1)^{-1}=(3, 1)$	(3, 1)	(3, 2, 1)	ϵ	(1, 2, 3)	(1, 2)	(2, 3)
$(1, 2)^{-1}=(1, 2)$	(1, 2)	(1, 2, 3)	(3, 2, 1)	ϵ	(2, 3)	(3, 1)
$(1, 2, 3)^{-1}=(3, 2, 1)$	(3, 2, 1)	(3, 1)	(1, 2)	(2, 3)	ϵ	(1, 2, 3)
$(3, 2, 1)^{-1}=(1, 2, 3)$	(1, 2, 3)	(1, 2)	(2, 3)	(3, 1)	(3, 2, 1)	ϵ

There are three partitions of 3, namely,

$$[3],\quad [2, 1],\quad [1^3].$$

There are therefore three classes of elements in $\mathfrak{S}_3$. Thus

$$C_{[3]}=(1, 2, 3)+(3, 2, 1),\quad C_{[2,1]}=(2, 3)+(3, 1)+(1, 2),\quad C_{[1^3]}=\epsilon.$$

Evidently

$$h_{[3]}=2,\quad h_{[2,1]}=3,\quad h_{[1^3]}=1.$$

The shapes associated with the three partitions are

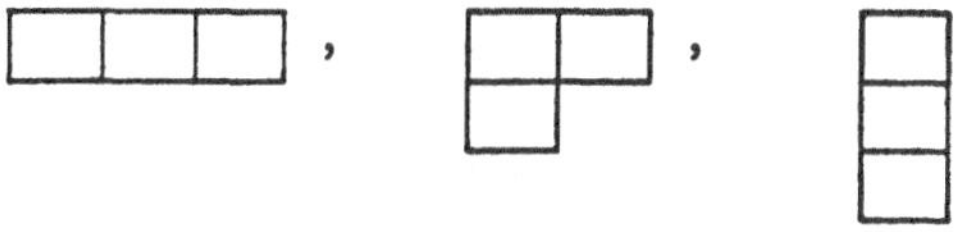

respectively and the corresponding values of f^{α} are

$$f^{[3]}=1,\quad f^{[2,1]}=2,\quad f^{[1^3]}=1.$$

Consequently, since $\theta^{\alpha} = 3!/f^{\alpha}$,

$$\theta^{[3]} = 6, \quad \theta^{[2,1]} = 3, \quad \theta^{[1^3]} = 6.$$

The standard tableaux are

$$S_1^{[3]} = \begin{array}{|c|c|c|}\hline 1 & 2 & 3 \\ \hline \end{array}\;;$$

$$S_1^{[2,1]} = \begin{array}{|c|c|}\hline 1 & 2 \\ \hline 3 & \multicolumn{1}{c}{} \\ \cline{1-1} \end{array}, \quad S_2^{[2,1]} = \begin{array}{|c|c|}\hline 1 & 3 \\ \hline 2 & \multicolumn{1}{c}{} \\ \cline{1-1} \end{array}\;;$$

$$S_1^{[1^3]} = \begin{array}{|c|}\hline 1 \\ \hline 2 \\ \hline 3 \\ \hline \end{array}\,.$$

The relations between the permutations and the semi-normal units are given by the following table :

	$e_{11}^{[3]}$	$e_{11}^{[2,1]}$	$e_{12}^{[2,1]}$	$e_{21}^{[2,1]}$	$e_{22}^{[2,1]}$	$e_{11}^{[1^3]}$	
ϵ	1	1	0	0	1	1	ϵ
(2, 3)	1	$-\frac{1}{2}$	$\frac{3}{4}$	1	$\frac{1}{2}$	-1	(2, 3)
(3, 1)	1	$-\frac{1}{2}$	$-\frac{3}{4}$	-1	$\frac{1}{2}$	-1	(3, 1)
(1, 2)	1	1	0	0	-1	-1	(1, 2)
(1, 2, 3)	1	$-\frac{1}{2}$	$\frac{3}{4}$	-1	$-\frac{1}{2}$	1	(3, 2, 1)
(3, 2, 1)	1	$-\frac{1}{2}$	$-\frac{3}{4}$	1	$-\frac{1}{2}$	1	(1, 2, 3)
	$6e_{11}^{[3]}$	$3e_{11}^{[2,1]}$	$3e_{21}^{[2,1]}$	$3e_{12}^{[2,1]}$	$3e_{22}^{[2,1]}$	$6e_{11}^{[1^3]}$	

This table can be read in two ways. For the formula

$$\tau_i = \sum_{\alpha, r, s} u_{rs\tau_i}^{\alpha} e_{rs}^{\alpha}$$

we ignore the column on the right and the bottom row ; thus

$$(3, 1) = e_{11}^{[3]} - \tfrac{1}{2}e_{11}^{[2,1]} - \tfrac{3}{4}e_{12}^{[2,1]} - e_{21}^{[2,1]} + \tfrac{1}{2}e_{22}^{[2,1]} - e_{11}^{[1^3]}.$$

For the formula

$$\theta^{\alpha} e_{rs}^{\alpha} = \sum_{\tau_i} u_{sr\tau_i}^{\alpha} \tau_i^{-1}$$

we ignore the column on the left and the top row ; thus

$$3e_{12}^{[2,1]} = (2, 3) - (3, 1) - (3, 2, 1) + (1, 2, 3).$$

The corresponding table for the natural units is

	$g_{11}^{[3]}$	$g_{11}^{[2,1]}$	$g_{12}^{[2,1]}$	$g_{21}^{[2,1]}$	$g_{22}^{[2,1]}$	$g_{11}^{[1^3]}$	
ϵ	1	1	0	0	1	1	ϵ
(2, 3)	1	0	1	1	0	−1	(2, 3)
(3, 1)	1	−1	0	−1	1	−1	(3, 1)
(1, 2)	1	1	−1	0	−1	−1	(1, 2)
(1, 2, 3)	1	−1	1	−1	0	1	(3, 2, 1)
(3, 2, 1)	1	0	−1	1	−1	1	(1, 2, 3)
	$6g_{11}^{[3]}$	$3g_{11}^{[2,1]}$	$3g_{21}^{[2,1]}$	$3g_{12}^{[2,1]}$	$3g_{22}^{[2,1]}$	$6g_{11}^{[1^3]}$	

By definition

$$T^{[3]} = e_{11}^{[3]} = g_{11}^{[3]},$$
$$T^{[2,1]} = e_{11}^{[2,1]} + e_{22}^{[2,1]} = g_{11}^{[2,1]} + g_{22}^{[2,1]},$$
$$T^{[1^3]} = e_{11}^{[1^3]} = g_{11}^{[1^3]}.$$

The relations between the expressions T^{α} and C_{β} are given by the following table, the centre of which is simply the table of group characters shown on page 70:

	$C_{[3]}$	$C_{[2,1]}$	$C_{[1^3]}$	
$6T^{[3]}$	1	1	1	$T^{[3]}$
$3T^{[2,1]}$	−1	0	2	$\frac{1}{2}T^{[2,1]}$
$6T^{[1^3]}$	1	−1	1	$T^{[1^3]}$
	$\frac{1}{2}C_{[3]}$	$\frac{1}{3}C_{[2,1]}$	$C_{[1^3]}$	

As before, this table can be read in two ways; thus

$$3T^{[2,1]} = -C_{[3]} + 2C_{[1^3]},$$
$$\tfrac{1}{3}C_{[2,1]} = T^{[3]} - T^{[1^3]}$$

BIBLIOGRAPHY

The following is a list of published writings which have some bearing on Substitutional Analysis. Those referred to in the text are given abbreviations which are enclosed in square brackets.

PAPERS

R. H. Bruck and T. L. Wade, " Bisymmetric tensor algebra Parts I and II ", *Amer. Journal of Math.* **64** (1942), pp. 725-753.

G. Frobenius, "Ueber die charakteristischen Einheiten der symmetrischen Gruppe", *Sitz. Preuss. Akad.* (1903), I, pp. 328-358. A number of other papers are listed in Speiser's *Theorie der Gruppen von endlicher Ordnung* (2nd edit.), p. 144.

D. E. Littlewood, " Group characters and the structure of groups ", *Proc. London Math. Soc.* (2), **39** (1935), pp. 150-199.

" On compound and induced matrices ", *Proc. London Math. Soc.* (2), **40** (1936), pp. 370-387.

" Polynomial concomitants and invariant matrices ", *Journal London Math. Soc.* **11** (1936), pp. 49-55.

D. E. Littlewood and A. R. Richardson, " Group characters and algebra ", *Phil. Trans. Roy. Soc.*, A, **233** (1934), pp. 99-141.

" Immanants of some special matrices ", *Quart. Journal of Math.* **5** (1934), pp. 269-282.

P. G. Molenaar, " Primitief-symmetrische projective invarianten I ", *Proc. Kon. Akad. Wet. Amsterdam,* **49** (1946), pp. 238-250.

" Primitief-symmetrische projective invarianten II ", *Proc. Kon. Akad. Wet. Amsterdam,* **49** (1946), pp. 357-368.

" Primitief-symmetrische projective invarianten III ", *Proc. Kon. Akad. Wet. Amsterdam,* **49** (1946), pp. 470-478.

F. D. Murnaghan, " On the representations of the symmetric group ", *Amer. Journal of Math.* **59** (1937), pp. 437-488. [Murnaghan.]

" The characters of the symmetric group ", *Amer. Journal of Math.* **59** (1937), pp. 739-753.

" The analysis of the direct product of irreducible representations of the symmetric groups ", *Amer. Journal of Math.* **60** (1938), pp. 44-65.

T. Nakayama, " On some modular properties of irreducible representations of a symmetric group ", *Japanese Journal of Math.* **17** (1940), pp. 165-184, 745-760. [Nakayama.]

G. de B. Robinson, " On the geometry of the linear representations of the symmetric group ", *Proc. London Math. Soc.* (2), **38** (1935), pp. 402-413.

" Note on an equation of quantitative substitutional analysis ", *Proc. London Math. Soc.* (2), **38** (1935), pp. 414-416.

" On the representation of the symmetric group ", *Amer. Journal of Math.* **60** (1938), pp. 745-760.

" On the representation of the symmetric group II ", *Amer. Journal of Math.* **69** (1947), pp. 286-298.

D. E. Rutherford, " On the relations between the numbers of standard tableaux ", *Proc. Edinburgh Math. Soc.* (2), **7** (1942), pp. 51-54. [Rutherford I.]

D. E. Rutherford, " On substitutional equations ", *Proc. Roy. Soc. Edinburgh*, A, **42** (1944), pp. 117-126. [Rutherford II.]

I. Schur, " Neue Begründung der Theorie der Gruppencharaktere ", *Sitz. Preuss. Akad.* (1905), pp. 406-432. [Schur.] A number of other papers are listed in Speiser's *Theorie der Gruppen von endlicher Ordnung* (2nd edit.), p. 145.

W. Specht, " Die irreduziblen Darstellungen der symmetrischen Gruppe ", *Math. Zeit.* **39** (1935), 696-711.

" Zur Darstellungstheorie der symmetrischen Gruppe ", *Math. Zeit.* **42** (1937), pp. 774-779.

R. M. Thrall, " Young's semi-normal representation of the symmetric group ", *Duke Math. Journal*, **8** (1941), pp. 611-624. [Thrall.]

T. L. Wade, " Tensor algebra and Young's symmetry operators ", *Amer. Journal of Math.* **63** (1941), pp. 645-657.

A. Young, " On quantitative substitutional analysis ", *Proc. London Math. Soc.* **33** (1900), pp. 97-146. [Young I.]

" On quantitative substitutional analysis (second paper) ", *Proc. London Math. Soc.* **34** (1902), pp. 361-397. [Young II.]

" On quantitative substitutional analysis (third paper) ", *Proc. London Math. Soc.* (2), **28** (1927), pp. 255-292. [Young III.]

" On quantitative substitutional analysis ", *Journal London Math. Soc.* **3** (1927), pp. 14-19.

" On quantitative substitutional analysis (fourth paper) ", *Proc. London Math. Soc.* (2), **31** (1929), pp. 253-272. [Young IV.]

" On quantitative substitutional analysis (fifth paper) ", *Proc. London Math. Soc.* (2), **31** (1929), pp. 273-288.

" On quantitative substitutional analysis (sixth paper) ", *Proc. London Math. Soc.* (2), **34** (1931), pp. 196-230. [Young VI.]

" Some generating functions ", *Proc. London Math. Soc.* (2), **35** (1931), pp. 425-444.

" On quantitative substitutional analysis (seventh paper) ", *Proc. London Math. Soc.* (2), **36** (1932), pp. 304-368.

" On quantitative substitutional analysis (eighth paper) ", *Proc. London Math. Soc.* (2), **37** (1933), pp. 441-495. [Young VIII.]

" The application of substitutional analysis to invariants ", *Phil. Trans. Roy. Soc.*, A, **234** (1935), pp. 79-114.

BOOKS

J. H. Grace and A. Young, *The algebra of invariants*. Cambridge, 1903.

D. E. Littlewood, *The theory of group characters*. Oxford, 1940. [Littlewood.]

P. G. Molenaar, *Eindige Substitutiegroepen* (Dissertation). Groningen, 1930.

F. D. Murnaghan, *The theory of group representations*. Baltimore, 1928.

B. L. van der Waerden, *Moderne Algebra II*. Berlin, 1931. [v. d. Waerden.]

H. Weyl, *Gruppentheorie und Quantenmechanik*. Leipzig, 1931.

The classical groups. Princeton, 1939.

The following books are also referred to in the text :

C. C. MacDuffee, *The theory of matrices*. Berlin, 1933. [MacDuffee.]

H. W. Turnbull and A. C. Aitken, *The theory of canonical matrices*. Glasgow, 1932. [T. A.]

INDEX